GENE CLONING

An introduction

Second edition

T. A. BROWN

UMIST, Manchester

CHAPMAN & HALL

University and Professional Division

London · New York · Tokyo · Melbourne · Madras

Published by Chapman & Hall, 2–6 Boundary Row, London SE1 8HN

Chapman & Hall, 2–6 Boundary Row, London SE1 8HN, UK

Chapman & Hall, 29 West 35th Street, New York NY10001, USA

Chapman & Hall Japan, Thomson Publishing Japan, Hirakawacho Nemoto Building, 7F, 1-7-11 Hirakawa-cho, Chiyoda-ku, Tokyo 102, Japan

Chapman & Hall Australia, Thomas Nelson Australia, 102 Dodds Street, South Melbourne, Victoria 3205, Australia

Chapman & Hall India, R. Seshadri, 32 Second Main Road, CIT East, Madras 600 035, India

First edition 1986
Reprinted 1987 (twice), 1988, 1989
Second edition 1990
Reprinted 1991

© 1990 T. A. Brown

Typeset in 10/11½ Palatino by Scarborough Typesetting Services, Scarborough
Printed in Great Britain by St Edmundsbury Press Ltd, Bury St Edmunds, Suffolk

ISBN 0 412 34210 3

A catalogue record for this book is available from the British Library

Library of Congress Cataloging-in-Publication data available

∞ Printed on permanent acid-free text paper, manufactured in accordance with the proposed ANSI/NISO Z 39.48–199X and ANSI Z 39.48–1984

Contents

Preface to the second edition

It was only when I started writing the Second Edition to this book that I fully appreciated how far gene cloning has progressed since 1986. Being caught up in the day to day excitement of biological research it is sometimes difficult to stand back and take a considered view of everything that is going on. The pace with which new techniques have been developed and applied to recombinant DNA research is quite remarkable. Procedures which in 1986 were new and innovative are now *de rigueur* for any self-respecting research laboratory and many of the standard techniques have found their way into undergraduate practical classes. Students are now faced with a vast array of different procedures for cloning genes and an even more diverse set of techniques for studying them once they have been cloned.

In revising this book I have tried to keep rigidly to a self-imposed rule that I would not make the Second Edition any more advanced than the first. There are any number of advanced texts for students or research workers who need detailed information on individual techniques and approaches. In contrast there is still a surprising paucity of really introductory texts on gene cloning. The First Edition was unashamedly introductory and I hope that the Second Edition will be also.

Nevertheless, changes were needed and on the whole the Second Edition contains more information. I have resisted the temptation to make many additions to Part One, where the fundamentals of gene cloning are covered. A few new vectors are described, especially for cloning in eukaryotes, but on the whole the first seven chapters are very much as they were in the First Edition. Part Two has been redefined so it now concentrates more fully on techniques for studying cloned genes, in particular with a description of methods for analysing gene regulation. Recombinant DNA techniques in general have become more numerous since 1986 and an undergraduate is now expected to have a broader appreciation of how cloned genes are studied. In Part Three the main theme is still biotechnology, but the tremendous advances in this area have required more extensive rewriting. The use of eukaryotes for synthesis of recombinant protein is now standard procedure, and we have seen the first great contributions of gene cloning to the study of human disease. The applications of gene cloning really make up a different book to this one, but nonetheless in Part Three I have tried to give a flavour of what is going on.

A number of people have been kind enough to comment on the First Edition and make suggestions for this revision. Don Grierson and Paul Sims again provided important and sensible advice. I must also thank Stephen Oliver and Richard Walmsley for their comments on specific parts of the book. Once again my wife's patience and encouragement has been a major factor in getting a Second Edition done at all. Finally I would like to thank all the students who have used the First Edition for the mainly nice things they have said about it.

T. A. Brown
Manchester

Preface to the first edition

This book is intended to introduce gene cloning and recombinant DNA technology to undergraduates who have no previous experience of the subject. As such, it assumes very little background knowledge on the part of the reader – just the fundamental details of DNA and genes that would be expected of an average sixth-former capable of a university entrance grade at A-level biology. I have tried to explain all the important concepts from first principles, to define all unfamiliar terms either in the text or in the glossary, to avoid the less-helpful jargon words, and to reinforce the text with as many figures as are commensurate with a book of reasonable price.

Although aimed specifically at first- and second-year undergraduates in biochemistry and related degree courses, I hope that this book will also prove useful to some experienced researchers. I have been struck over the last few years by the number of biologists, expert in other aspects of the science, who have realized that gene cloning may have a role in their own research projects. Possibly this text can act as a painless introduction to the complexities of recombinant DNA technology for those of my colleagues wishing to branch out into this new discipline.

I would like to make it clear that this book is not intended as competition for the two excellent gene cloning texts already on the market. I have considerable regard for the books by Drs Old and Primrose and by Professor Glover, but believe that both texts are aimed primarily at advanced undergraduates who have had some previous exposure to the subject. It is this 'previous exposure' that I aim to provide. My greatest satisfaction will come if this book is accepted as a primer for Old and Primrose or for Glover.

I underestimated the effort needed to produce such a book and must thank several people for their help. The publishers provided the initial push to get the project under way. I am indebted to Don Grierson at Nottingham University and Paul Sims at UMIST for reading the text and suggesting improvements; all errors and naïveties are, however, mine. Finally, my wife Keri typed most of the manuscript and came to my rescue on several occasions with the right word or turn of phrase. This would never have been finished without her encouragement.

T. A. Brown
Manchester

Part One

The Basic Principles of Gene Cloning

Why gene cloning is important

1

Just over a century ago, Gregor Mendel formulated a set of rules to explain the inheritance of biological characteristics. The basic assumption of these rules is that each heritable property of an organism is controlled by a factor, called a **gene**, that is a physical particle present somewhere in the cell. The rediscovery of Mendel's laws in 1900 marks the birth of **genetics**, the science aimed at understanding what these genes are and exactly how they work.

1.1 THE EARLY DEVELOPMENT OF GENETICS

For the first 30 years of its life this new science grew at an astonishing rate. The idea that genes reside on **chromosomes** was proposed by W. Sutton in 1903, and received experimental backing from T. H. Morgan in 1910. Morgan and his colleagues then developed the techniques for **gene mapping**, and by 1922 had produced a comprehensive analysis of the relative positions of over 2000 genes on the four chromosomes of the fruit fly, *Drosophila melanogaster*.

Despite the brilliance of these classical genetic studies, there was no real understanding of the molecular nature of the gene until the 1940s. Indeed, it was not until the experiments of Avery, MacLeod and McCarty in 1944, and of Hershey and Chase in 1952, that anyone believed DNA to be the genetic material; up to then it was widely thought that genes were made of protein. The discovery of the role of DNA was a tremendous stimulus to genetic research, and many famous biologists (Delbruck, Chargaff, Crick and Monod were among the most influential) contributed to the second great age of genetics. In the 14 years between 1952 and 1966 the structure of DNA was elucidated, the genetic code cracked, and the processes of transcription and translation described.

1.2 THE ADVENT OF GENE CLONING

These years of activity and discovery were followed by a lull, a period of anticlimax when it seemed to some molecular biologists (as the new

generation of geneticists styled themselves) that there was little of fundamental importance that was not understood. In truth there was a frustration
that the experimental techniques of the late 1960s were not sophisticated
enough to allow the gene to be studied in any greater detail.

Then, in the years 1971–73 genetic research was thrown back into gear by
what can only be described as a revolution in modern biology. A whole new
methodology was developed, allowing previously impossible experiments
to be planned and carried out, if not with ease, then at least with success.
These methods, referred to as **recombinant DNA technology** or **genetic
engineering**, and having at their core the process of **gene cloning**, sparked
the third great age of genetics. We are still in the midst of the boom caused by
this revolution and there is no end to the excitement in sight.

1.3 WHAT IS GENE CLONING?

The basic steps in a gene cloning experiment are as follows (Figure 1.1).

1. A fragment of DNA, containing the gene to be cloned, is inserted into a
   circular DNA molecule called a **vector**, to produce a **chimaera** or
   recombinant DNA molecule.
2. The vector acts as a **vehicle** that transports the gene into a host cell, which
   is usually a bacterium, although other types of living cell can be used.
3. Within the host cell the vector multiplies, producing numerous identical
   copies not only of itself but also of the gene that it carries.
4. When the host cell divides, copies of the recombinant DNA molecule are
   passed to the progeny and further vector replication takes place.
5. After a large number of cell divisions, a colony, or **clone**, of identical host
   cells is produced. Each cell in the clone contains one or more copies of the
   recombinant DNA molecule; the gene carried by the recombinant
   molecule is now said to be cloned.

1.4 GENE CLONING REQUIRES SPECIALIZED TOOLS AND TECHNIQUES

1.4.1 VEHICLES

The central component of a gene cloning experiment is the vehicle, which
transports the gene into the host cell and is responsible for its replication. To
act as a cloning vehicle a DNA molecule must be capable of entering a host
cell and, once inside, replicating to produce multiple copies of itself. Two
naturally occurring types of DNA molecule satisfy these requirements:

1. **Plasmids**, which are small circles of DNA found in bacteria and some
   other organisms. Plasmids can replicate independently of the host cell
   chromosome.

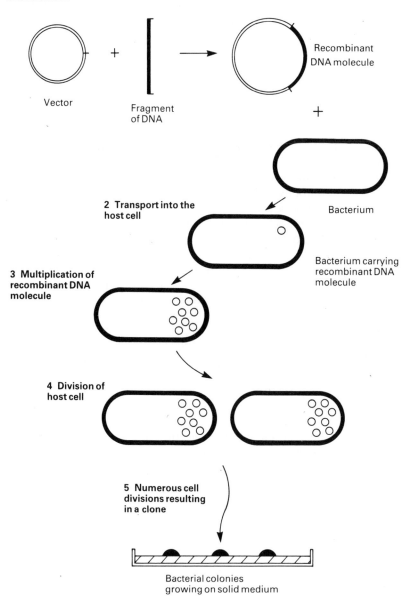

Figure 1.1 The basic steps in gene cloning.

2. **Virus chromosomes**, in particular the chromosomes of **bacteriophages**, which are viruses that specifically infect bacteria. During infection the bacteriophage DNA molecule is injected into the host cell where it undergoes replication.

Chapter 2 covers the basic features of plasmids and bacteriophage chromosomes, providing the necessary background for an understanding of how these molecules are used as cloning vehicles.

1.4.2 TECHNIQUES FOR HANDLING DNA

Plasmids and bacteriophage DNA molecules display the basic properties required of potential cloning vehicles. But this potential would be wasted without experimental techniques for handling DNA molecules in the laboratory. The fundamental steps in gene cloning, as described on p. 4 and in Figure 1.1, require several manipulative skills (Table 1.1). First, pure samples of DNA must be available, both of the cloning vehicle and of the gene to be cloned. The methods used to purify DNA from living cells are outlined in Chapter 3.

Having prepared samples of DNA, construction of a recombinant DNA molecule requires that the vector be cut at a specific point and then repaired in such a way that the gene is inserted into the vehicle. The ability to manipulate DNA in this way is an offshoot of basic research into DNA synthesis and modification within living cells. The discovery of enzymes that can cut or join DNA molecules in the cell has led to the purification of **restriction endonucleases** and **ligases**, which are now used to construct recombinant DNA molecules in the test-tube. The properties of these enzymes, and the way they are used in gene cloning experiments, are described in Chapter 4.

Once a recombinant DNA molecule has been constructed, it must be introduced into the host cell so that replication can take place. Transport into the host cell makes use of natural processes for uptake of plasmid and viral DNA molecules. These processes, and the ways they are utilized in gene cloning, are described in Chapter 5.

Table 1.1 Basic skills needed to carry out a simple gene cloning experiment

1. Preparation of pure samples of DNA	(Chapter 3)
2. Cutting DNA molecules	(Chapter 4, pp. 56/65)
3. Analysis of DNA fragment sizes	(Chapter 4, pp. 65/72)
4. Joining DNA molecules together	(Chapter 4, pp. 73/83)
5. Introduction of DNA into host cells	(Chapter 5, pp. 84/90)
6. Identification of cells that contain recombinant DNA molecules	(Chapter 5, pp. 90/100, and Chapter 8)

1.4.3 THE DIVERSITY OF CLONING VECTORS

Although gene cloning is relatively new, it has nevertheless developed into a very sophisticated technology. Today a wide variety of different cloning vectors are available. All are derived from naturally occurring plasmids or viruses, but most have been modified in various ways to suit each for a particular type of cloning experiment. In Chapters 6 and 7 the most important types of vector are described, and their uses examined.

1.5 WHY GENE CLONING IS SO IMPORTANT

As you can see from Figure 1.1, gene cloning is a relatively straightforward procedure. Why then has it assumed such importance in biology? The answer is largely because cloning can provide a pure sample of an individual gene, separated from all the other genes that it normally shares the cell with. To understand exactly how this arises, consider a gene cloning experiment drawn in a slightly different way (Figure 1.2). The DNA fragment to be cloned can be part of a mixture of many different fragments, each carrying a different gene or part of a gene. This mixture could indeed be the entire genetic complement of an organism, man for instance. All these fragments will be inserted into different vector molecules to produce a family of recombinant DNA molecules, one of which carries the gene of interest. Usually only one recombinant DNA molecule will be transported into any single host cell, so that although the final set of clones may contain many different recombinant DNA molecules, each individual clone will contain multiple copies of just one molecule. The gene is now separated away from all the other genes in the original mixture, and its specific features can be studied in detail.

In practice, the key to the success or failure of a cloning experiment is the ability to identify the particular clone of interest from the many different ones that are obtained. If we consider the **genome** of the bacterium *Escherichia coli*, which contains something in the region of 2000 different genes, we might at first despair of being able to find just one gene amongst all the possible clones (Figure 1.3). The problem becomes even more overwhelming when we remember that bacteria are relatively simple organisms and that the human genome contains at least 20 times as many genes. However, as explained in Chapter 8, a variety of different strategies can be used to ensure that the correct gene can be obtained at the end of the cloning experiment. Some of these strategies involve modifications to the basic cloning procedure, so that only cells containing the desired recombinant DNA molecule can divide and the clone of interest is automatically **selected**. Other methods involve techniques that enable the desired clone to be identified from a mixture of lots of different clones.

Once a gene has been cloned there is almost no limit to the information that can be obtained about the structure and expression of that gene. The availability of cloned material has stimulated the development of analytical

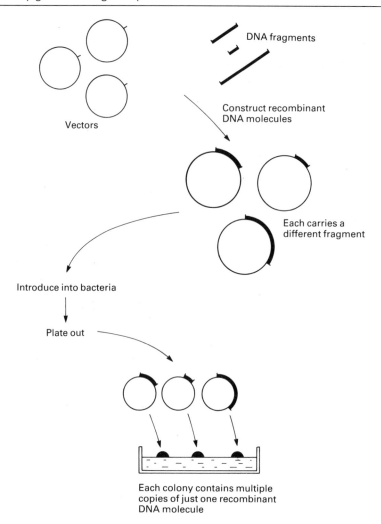

Figure 1.2 Cloning allows individual fragments of DNA to be purified.

methods for studying genes, with new techniques being introduced all the time. Methods for studying the structure and expression of a cloned gene are described in Chapters 9 and 10 respectively.

1.6 THE IMPACT OF CLONING ON RESEARCH AND BIOTECHNOLOGY

There are very few areas of biological research that have not been touched by gene cloning. In industry, for example, the ability to clone genes has led to

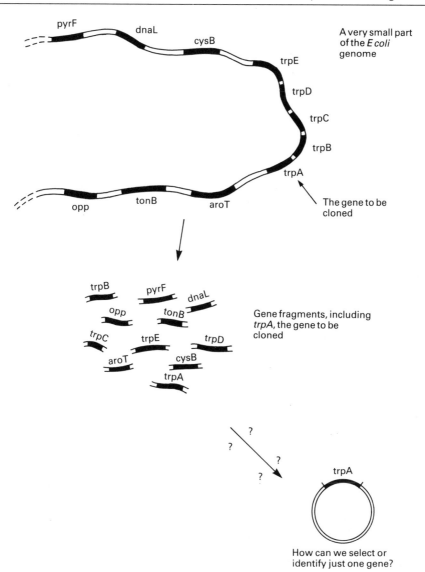

Figure 1.3 The problem of selection.

far-reaching advances in **biotechnology**. For many years man has made use of microorganisms such as bacteria and fungi as living factories in which to produce useful compounds. Examples are provided by antibiotics, such as penicillin, which is synthesized by a fungus called *Penicillium*, and strepto-mycin, produced by the bacterium *Streptomyces griseus*. Gene cloning has revolutionized biotechnology, most notably in providing a way in which

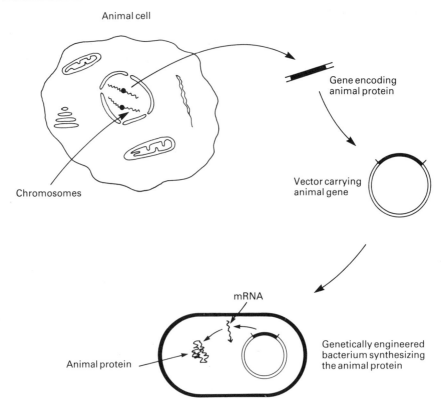

Figure 1.4 A possible scheme for the production of an animal protein by a bacterium.

mammalian proteins can be produced in bacterial cells. A remarkable property of a cloned gene is that it can often be made to function in an organism totally unrelated to that in which it is normally found. For example, an animal gene can be transferred by cloning into a bacterium and then induced by some careful modifications to carry on working as though nothing had happened (Figure 1.4). The implications are enormous. Genes controlling the synthesis of important pharmaceuticals, such as drugs and hormones, can be taken from the organism in which they occur naturally, but from which they may be costly and difficult to prepare, and placed in a bacterium or other type of organism, from which the product can be recovered conveniently and in large quantities. A number of successes have been notched up by biotechnologists, with recombinant insulin being the most noteworthy achievement so far. Production of protein by cloned genes is described in Chapters 11 and 12.

Despite the achievements of biotechnology, the most important impact of gene cloning has been in medical research. New types of vaccines, providing protection against diseases for which vaccination was previously impossible, have been developed thanks to the ability to clone genes. Many inherited diseases can now be diagnosed in an unborn child, and recent research has led to the hope that cystic fibrosis and other heart-rending afflictions will soon be treatable. The way in which gene cloning is being applied in medical research is described in the final chapter of this book.

FURTHER READING

Cherfas, J. (1982) *Man Made Life*, Blackwell, Oxford, UK – a history of genetic engineering.

Portugal, F. H. and Cohen, J. S. (1977) *A Century of DNA*, MIT Press, Cambridge, USA – charts the development of genetics and molecular biology.

Felsenfeld, G. (1985) DNA. *Scientific American*, **253**, (Oct): 44–53 – a brief review of the important features of DNA.

Brown, T. A. (1989) *Genetics: A Molecular Approach*, Van Nostrand Reinhold, London – an introduction to genetics and molecular biology.

Vehicles: plasmids and bacteriophages

2

A DNA molecule needs to display several features to be able to act as a vehicle for gene cloning. Most important, it must be able to replicate within the host cell, so that numerous copies of the recombinant DNA molecule can be produced and passed to the daughter cells. A cloning vehicle also needs to be relatively small, ideally less than 10 kilobases (kb) in size, as large molecules tend to break down during purification, and are also more difficult to manipulate. Two kinds of DNA molecule that satisfy these criteria can be found in bacterial cells: plasmids and bacteriophage chromosomes. Although plasmids are frequently employed as cloning vehicles, two of the most important types of vector in use today are derived from bacteriophages.

2.1 PLASMIDS

2.1.1 BASIC FEATURES OF PLASMIDS

Plasmids are circular molecules of DNA that lead an independent existence in the bacterial cell (Figure 2.1). Plasmids almost always carry one or more genes, and often these genes are responsible for a useful characteristic displayed by the host bacterium. For example, the ability to survive in normally toxic concentrations of antibiotics such as chloramphenicol or ampicillin is often due to the presence in the bacterium of a plasmid carrying antibiotic-resistance genes. In the laboratory antibiotic resistance is often used as a **selectable marker** to ensure that bacteria in a culture contain a particular plasmid (Figure 2.2).

All plasmids possess at least one DNA sequence that can act as an **origin of replication**, so they are able to multiply within the cell quite independently of the main bacterial chromosome (Figure 2.3a). The smaller plasmids make use of the host cell's own DNA replicative enzymes in order to make copies of themselves, whereas some of the larger ones carry genes that code for special enzymes that are specific for plasmid replication.

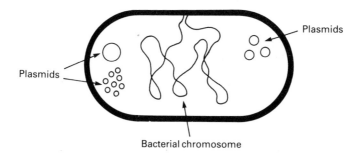

Figure 2.1 Plasmids: independent genetic elements found in bacterial cells.

A few types of plasmid are also able to replicate by inserting themselves into the bacterial chromosome (Figure 2.3b). These integrative plasmids or **episomes** may be stably maintained in this form through numerous cell divisions, but will at some stage exist as independent elements. Integration is also an important feature of some bacteriophage chromosomes and will be described in more detail when these are considered (p. 18).

2.1.2 SIZE AND COPY NUMBER

These two features of plasmids are particularly important as far as cloning is concerned. We have already mentioned the relevance of plasmid size and stated that less than 10 kb is desirable for a cloning vehicle. Plasmids range from about 1.0 kb for the smallest to over 250 kb for the largest plasmids (Table 2.1), so only a few will be useful for cloning purposes. However, as described in Chapter 7, larger plasmids may be adapted for cloning under some circumstances.

The **copy number** refers to the number of molecules of an individual plasmid that are normally found in a single bacterial cell. The factors that

Table 2.1 Sizes of representative plasmids

Plasmid	Size		Organism
	Nucleotide length (kb)	Molecular wt (MDa)	
pBR345	0.7	0.46	*E. coli*
pBR322	4.362	2.9	*E. coli*
ColEl	6.36	4.2	*E. coli*
RP4	54	36	*Pseudomonas* + others
F	95	63	*E. coli*
TOL	117	78	*Pseudomonas putida*
pTiAch5	213	142	*Agrobacterium tumefaciens*

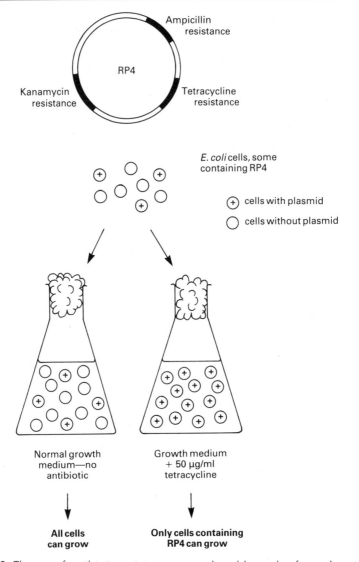

Figure 2.2 The use of antibiotic resistance as a selectable marker for a plasmid. RP4 (top) carries genes for resistance to ampicillin, tetracycline and kanamycin. Only those *E. coli* cells that contain RP4 (or a related plasmid) will be able to survive and grow in a medium that contains toxic amounts of one or more of these antibiotics.

control copy number are not well understood, but each plasmid has a characteristic value that may be as low as one (especially for the large molecules) or as many as 50 or more. Generally speaking, a useful cloning vehicle needs to be present in the cell in multiple copies so that large quantities of the recombinant DNA molecule can be obtained.

(a) **Non-integrative plasmid**

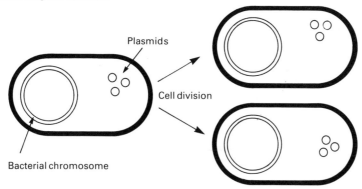

(b) **Episome**

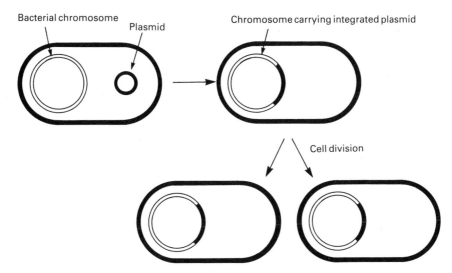

Figure 2.3 Replication strategies for (a) a non-integrative plasmid, and (b) an episome.

2.1.3 CONJUGATION AND COMPATIBILITY

Plasmids fall into two groups: conjugative and non-conjugative. Conjugative plasmids are characterized by the ability to promote sexual **conjugation** between bacterial cells (Figure 2.4), a process that can result in a conjugative plasmid spreading from one cell to all the other cells in a bacterial culture. Conjugation and plasmid transfer are controlled by a set of transfer or *tra* genes, which are present on conjugative plasmids but absent

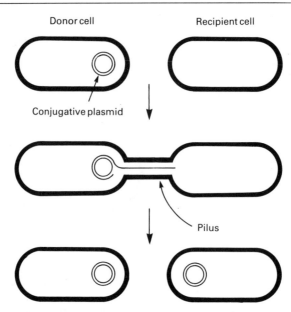

Figure 2.4 Plasmid transfer by conjugation between bacterial cells. The donor and recipient cells attach to each other by a pilus, an appendage present on the surface of the donor cell. A copy of the plasmid is then passed through the pilus to the recipient cell.

from the non-conjugative type. However, a non-conjugative plasmid may, under some circumstances, be cotransferred along with a conjugative plasmid when both are present in the same cell.

Several different kinds of plasmid may be found in a single cell, including more than one different conjugative plasmid at any one time. In fact, cells of *E. coli* have been known to contain up to seven different plasmids at once. To be able to coexist in the same cell, different plasmids must be **compatible**. If two plasmids are incompatible then one or the other will be quite rapidly lost from the cell. Different types of plasmid can therefore be assigned to different **incompatibility groups** on the basis of whether or not they can coexist, and plasmids from a single incompatibility group are often related to each other in various ways. The basis of incompatibility is not well understood, but events during plasmid replication are thought to underlie the phenomenon.

2.1.4 PLASMID CLASSIFICATION

The most useful classification of naturally occurring plasmids is based on the main characteristic coded by the plasmid genes. The five main types of plasmid according to this classification are as follows.

1. **Fertility** or **'F' plasmids** carry only *tra* genes and have no characteristic beyond the ability to promote conjugal transfer of plasmids; e.g. F plasmid of *E. coli*.
2. **Resistance** or **'R' plasmids** carry genes conferring on the host bacterium resistance to one or more antibacterial agents, such as chloramphenicol, ampicillin and mercury. R plasmids are very important in clinical microbiology as their spread through natural populations can have profound consequences in the treatment of bacterial infections; e.g. RP4, commonly found in *Pseudomonas*, but also occurring in many other bacteria.
3. **Col plasmids** code for colicins – proteins that kill other bacteria; e.g. ColE1 of *E coli*.
4. **Degradative plasmids** allow the host bacterium to metabolize unusual molecules such as toluene and salicylic acid; e.g. TOL (= pWWO) of *Pseudomonas putida*.
5. **Virulence plasmids** confer pathogenicity on the host bacterium; e.g. **Ti plasmids** of *Agrobacterium tumefaciens*, which induce crown gall disease on dicotyledonous plants.

2.1.5 PLASMIDS IN ORGANISMS OTHER THAN BACTERIA

Although plasmids are widespread in bacteria they are by no means so common in other organisms. The best characterized eukaryotic plasmid is the **2 μm circle** that occurs in many strains of the yeast *Saccharomyces cerevisiae*. The discovery of the 2 μm plasmid was very fortuitous as it has allowed the construction of vectors for cloning genes with this very important industrial organism as the host (p. 128). However, the search for plasmids in other eukaryotes (e.g. filamentous fungi, plants and animals) has proved disappointing, and it is suspected that many higher organisms simply do not harbour plasmids within their cells.

2.2 BACTERIOPHAGES

2.2.1 BASIC FEATURES OF BACTERIOPHAGES

Bacteriophages, or phages as they are commonly known, are viruses that specifically infect bacteria. Like all viruses, phages are very simple in structure, consisting merely of a DNA (or occasionally RNA) molecule carrying a number of genes, including several for replication of the phage, surrounded by a protective coat or **capsid** made up of protein molecules (Figure 2.5).

The general pattern of infection, which is the same for all types of phage, is a three-step process (Figure 2.6).

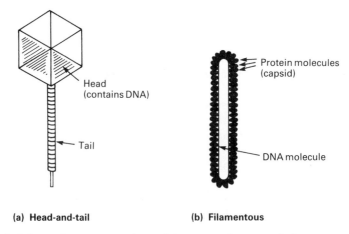

(a) Head-and-tail   (b) Filamentous

Figure 2.5 Schematic representations of the two main types of phage structure. (a) Head-and-tail (e.g. λ). (b) Filamentous (e.g. M13).

1. The phage particle attaches to the outside of the bacterium and injects its DNA chromosome into the cell.
2. The phage DNA molecule is replicated, usually by specific phage enzymes coded by genes on the phage chromosome.
3. Other phage genes direct synthesis of the protein components of the capsid, and new phage particles are assembled and released from the bacterium.

With some phage types the entire infection cycle is completed very quickly, possibly in less than 20 minutes. This type of rapid infection is called a **lytic cycle**, as release of the new phage particles is associated with lysis of the bacterial cell. The characteristic feature of a lytic infection cycle is that phage DNA replication is immediately followed by synthesis of capsid proteins, and the phage DNA molecule is never maintained in a stable condition in the host cell.

2.2.2 LYSOGENIC PHAGES

In contrast to a lytic cycle, **lysogenic** infection is characterized by retention of the phage DNA molecule in the host bacterium, possibly for many thousands of cell divisions. With many lysogenic phages the phage DNA is inserted into the bacterial genome, in a manner similar to episomal insertion (see Figure 2.3). The integrated form of the phage DNA (called the **prophage**) is quiescent, and a bacterium (referred to as a **lysogen**) which carries a prophage is usually physiologically indistinguishable from an uninfected cell. However, the prophage is eventually released from the host

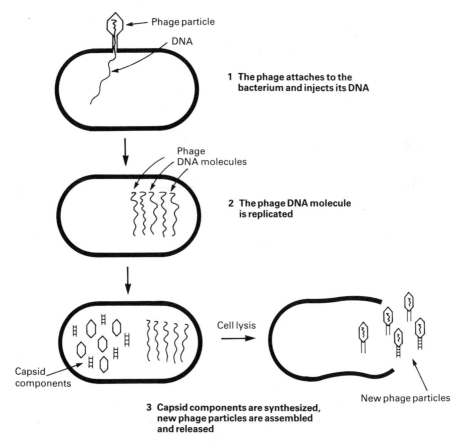

Figure 2.6 The general pattern of infection of a bacterial cell by a bacteriophage.

genome and the phage reverts to the lytic mode and lyses the cell. The infection cycle of λ, a typical lysogenic phage of this type, is shown in Figure 2.7.

A limited number of lysogenic phages follow a rather different infection cycle. When **M13,** or a related phage, infects *E. coli*, new phage particles are continuously assembled and released from the cell. The M13 DNA is not integrated into the bacterial genome and does not become quiescent. With these phages, cell lysis never occurs, and the infected bacterium can continue to grow and divide, albeit at a slower rate than uninfected cells. Figure 2.8 shows the M13 infection cycle.

Although there are many different varieties of bacteriophage, only λ and M13 have found any real role as cloning vectors. The properties of these two phages will now be considered in more detail.

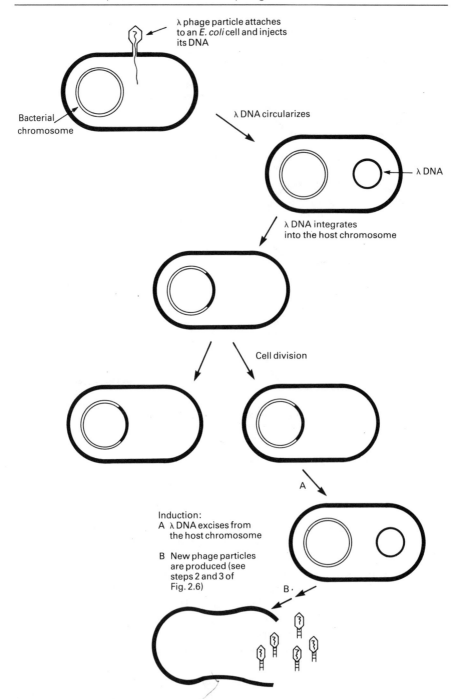

Figure 2.7 The lysogenic infection cycle of bacteriophage λ.

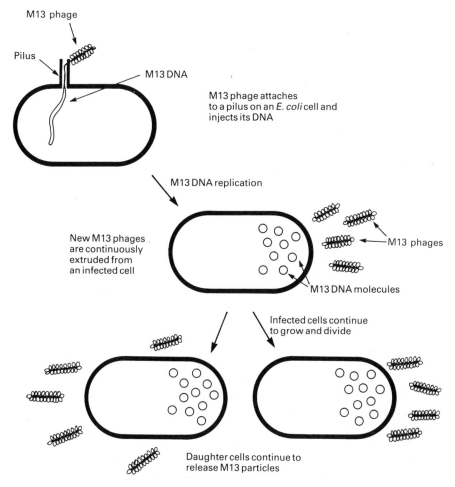

Figure 2.8 The infection cycle of bacteriophage M13.

(a) Gene organization in the λ DNA molecule

λ is a typical example of a head-and-tail phage (Figure 2.5a). The DNA is contained in the polyhedral head structure and the tail serves to attach the phage to the bacterial surface and to inject the DNA into the cell (Figure 2.7).

The λ DNA molecule is 49 kb in size and has been intensively studied by the techniques of gene mapping and **DNA sequencing**. As a result the positions and identities of most of the genes on the λ DNA molecule are known (Figure 2.9). A feature of the λ genetic map is that genes related in terms of function are clustered together on the genome. For example, all of the genes coding for components of the capsid are grouped together in the lefthand third of the molecule, and genes controlling integration of the

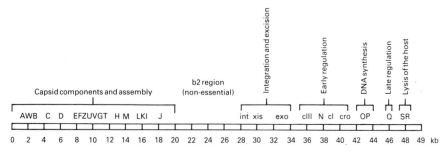

Figure 2.9 The λ genetic map, showing the positions of the important genes and the functions of the gene clusters.

prophage into the host genome are clustered in the middle of the molecule. Clustering of related genes is profoundly important for controlling expression of the λ genome, as it allows genes to be switched on and off as a group rather than individually. Clustering is also important in the construction of λ-based cloning vectors (described in Chapter 6).

(b) The linear and circular forms of λ DNA

A second feature of λ that turns out to be of importance in the construction of cloning vectors is the conformation of the DNA molecule. The molecule shown in Figure 2.9 is linear, with two free ends, and represents the DNA present in the phage head structure. This linear molecule consists of two **complementary** strands of DNA, base-paired according to the **Watson–Crick rules** (that is, double-stranded DNA). However, at either end of the molecule is a short 12 nucleotide stretch, in which the DNA is single-stranded (Figure 2.10a). The two single strands are complementary, and so can base-pair with one another to form a circular, completely double-stranded molecule (Figure 2.10b).

Complementary single strands are often referred to as '**sticky**' ends or 'cohesive' ends, because base-pairing between them can 'stick' together the two ends of a DNA molecule (or the ends of two different DNA molecules). The λ cohesive ends are called the ***cos* sites** and they play two distinct roles during the λ infection cycle. First of all, they allow the linear DNA molecule that is injected into the cell to be circularized, which is a necessary prerequisite for insertion into the bacterial genome (Figure 2.7).

The second role of the *cos* sites is rather different, and comes into play after the prophage has excised from the host genome. At this stage a large number of new λ DNA molecules are produced by the rolling circle mechanism of replication (Figure 2.10c), in which a continuous DNA strand is 'rolled off' of the template molecule. The result is a catenane consisting of a series of complete, linear λ genomes joined together at the *cos* sites. The role of the *cos* sites is now to act as recognition sequences for an **endonuclease**

(a) The linear form of the λ DNA molecule

Left cohesive end

Right cohesive end

CCGCCGCTGGA

G GGC GGCG ACC T

(b) The circular form of the λ DNA molecule

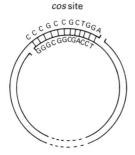

cos site

(c) Replication and packaging of λ DNA

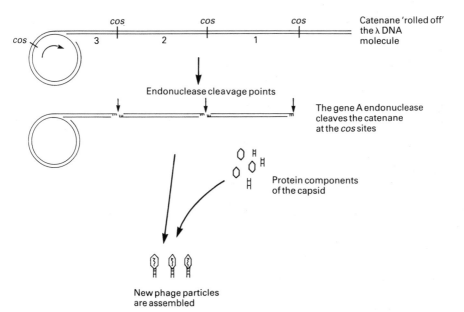

cos   cos   cos   Catenane 'rolled off' the λ DNA molecule

cos   3   2   1

Endonuclease cleavage points

The gene A endonuclease cleaves the catenane at the cos sites

Protein components of the capsid

New phage particles are assembled

Figure 2.10 The linear and circular forms of λ DNA. (a) The linear form showing the left and right cohesive ends. (b) Base pairing between the cohesive ends results in the circular form of the molecule. (c) Rolling circle replication produces a catenane of new linear λ DNA molecules, which are individually packaged into phage heads as new λ particles are assembled.

which cleaves the catenane at the *cos* sites producing complete single λ genomes. This endonuclease (which is the product of gene A on the λ DNA molecule) creates the single-stranded sticky ends, and also acts in conjunction with other proteins to package the λ genomes into the phage head structure.

As we shall see in Chapter 6, the cleavage and packaging processes recognize just the *cos* sites and the DNA sequences to either side of them. Changing the structure of the internal regions of the λ genome, for example by inserting new genes, has no effect on these events so long as the overall length of the λ genome is not altered too greatly.

(c) M13 – a filamentous phage

M13 is an example of a filamentous phage (Figure 2.5b) and is completely different in structure from λ. Furthermore, the M13 DNA molecule is much smaller than the λ genome, being only 6407 nucleotides in length. It is circular, and is unusual in that it consists entirely of single-stranded DNA.

The smaller size of the M13 DNA molecule means that it has room for fewer genes than the λ genome. This is possible because the M13 capsid is constructed from multiple copies of just three proteins (requiring only three genes), whereas synthesis of the λ head-and-tail structure involves over 15 different proteins. In addition, M13 follows a simpler infection cycle than λ and does not need genes for insertion into the host genome.

Injection of an M13 DNA molecule into an *E. coli* cell occurs via the **pilus**, the structure that connects two cells during sexual conjugation (see Figure 2.4). Once inside the cell the single-stranded molecule acts as the template for synthesis of a complementary strand, resulting in normal double-stranded DNA (Figure 2.11a). This molecule is not inserted into the bacterial genome, but instead replicates until over 100 copies are present in the cell (Figure 2.11b). When the bacterium divides, each daughter receives copies of the phage genome, which continues to replicate, thereby maintaining its overall numbers per cell. As shown in Figure 2.11c, new phage particles are continuously assembled and released, about 1000 new phages being produced during each generation of an infected cell.

(d) The attraction of M13 as a cloning vehicle

Several features of M13 make this phage attractive as the basis for a cloning vehicle. The genome is less than 10 kb in size, well within the range that we stated was desirable for a potential vector. In addition, the double-stranded **replicative form** of the M13 genome behaves very much like a plasmid and can be treated as such for experimental purposes. It is easily prepared from a culture of infected *E. coli* cells (p. 48) and can be reintroduced by **transfection** (p. 95).

Most importantly, genes cloned with an M13-based vector can be obtained in the form of single-stranded DNA. Single-stranded versions of

(a) Injection of single-stranded DNA into the host cell, followed by synthesis of the second strand

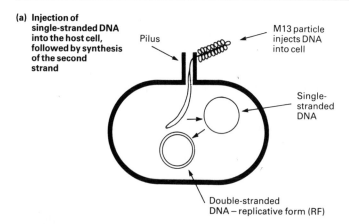

Pilus

M13 particle injects DNA into cell

Single-stranded DNA

Double-stranded DNA – replicative form (RF)

(b) Replication of the RF to produce new double-stranded molecules

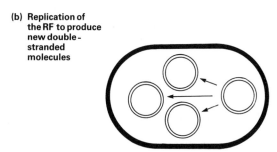

(c) Mature M13 phage are continuously produced

RF replicates by rolling circle mechanism to produce linear single-stranded DNA

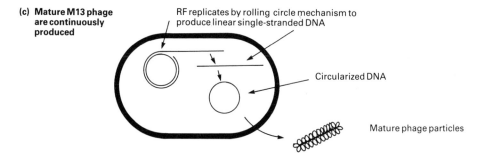

Circularized DNA

Mature phage particles

Figure 2.11 The M13 infection cycle showing the different types of DNA replication that occur. (a) After infection the single-stranded M13 DNA molecule is converted into the double-stranded replicative form (RF). (b) The RF replicates to produce multiple copies of itself. (c) Single-stranded molecules are synthesized by rolling circle replication and used in the assembly of new M13 particles.

cloned genes are needed for several techniques, notably DNA sequencing and *in vitro* mutagenesis (pp. 186 and 215). Using an M13 vector is an easy and reliable way of obtaining single-stranded DNA for this type of work.

2.2.3 VIRUSES AS CLONING VEHICLES FOR OTHER ORGANISMS

Most living organisms are infected by viruses and it is not surprising that great interest has been shown in the possibility that viruses may be used as cloning vehicles for higher organisms. This is especially important when it is remembered that plasmids are not commonly found in organisms other than bacteria and yeast (p. 17).

In fact viruses have considerable potential as cloning vehicles for both animals and plants. Mammalian viruses such as **simian virus 40 (SV40)** and plant viruses such as **cauliflower mosaic virus (CaMV)** are the ones that have received most attention so far, but several others are also being studied. These are discussed more fully in Chapter 7.

FURTHER READING

Hardy, K. (1986) *Bacterial Plasmids*, 2nd edn, Van Nostrand Reinhold, Wokingham, UK.

Lin, E. C. C. *et al.* (1984) *Bacteria, Plasmids and Phages*, Harvard University Press, Cambridge, USA.

Freifelder, D. (1987) *Microbial Genetics*, Jones and Bartlett, Boston, USA.

Each of these books provides a detailed description of the molecules used as the basis for most cloning vectors.

Purification of DNA from living cells

3

The genetic engineer will, at different times, need to prepare at least three distinct kinds of DNA. Firstly, **total cell DNA** will often be required as a source of material from which to obtain genes to be cloned. Total cell DNA may be DNA from a culture of bacteria, from a plant, from animal cells, or from any other type of organism that is being studied.

The second type of DNA that will be required is pure plasmid DNA. Preparation of plasmid DNA from a culture of bacteria follows the same basic steps as purification of total cell DNA, with the crucial difference that at some stage the plasmid DNA must be separated from the main bulk of chromosomal DNA also present in the cell.

Finally, phage DNA will be needed if a phage cloning vehicle is to be used. Phage DNA is generally prepared from bacteriophage particles rather than from infected cells, so there is no problem with contaminating bacterial DNA. However, special techniques are needed to remove the phage capsid. An exception is the double-stranded replicative form of M13 which is prepared from *E. coli* cells just as though it were a bacterial plasmid.

3.1 PREPARATION OF TOTAL CELL DNA

The fundamentals of DNA preparation are most easily understood by first considering the simplest type of DNA purification procedure, that where the entire DNA complement of a bacterial cell is required. The modifications needed for plasmid and phage DNA preparation can then be described later.

The procedure for total DNA preparation from a culture of bacterial cells can be divided into four stages (Figure 3.1).

1. A culture of bacteria is grown and then **harvested**.
2. The cells are broken open to release their contents.
3. This **cell extract** is treated to remove all components except the DNA.
4. The resulting DNA solution is concentrated.

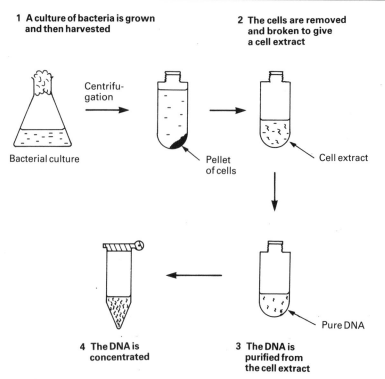

1 **A culture of bacteria is grown and then harvested**

2 **The cells are removed and broken to give a cell extract**

Centrifugation

Bacterial culture

Pellet of cells

Cell extract

Pure DNA

4 **The DNA is concentrated**

3 **The DNA is purified from the cell extract**

Figure 3.1 The basic steps in preparation of total cell DNA from a culture of bacteria.

3.1.1 GROWING AND HARVESTING A BACTERIAL CULTURE

Most bacteria can be grown without too much difficulty in a liquid medium (**broth culture**). The culture medium must provide a balanced mixture of the essential nutrients at concentrations that will allow the bacteria to grow and divide efficiently. Two typical growth media are detailed in Table 3.1. M9 is an example of a **defined medium** in which all the components are known. This medium contains a mixture of inorganic nutrients to provide essential elements such as nitrogen, magnesium and calcium, as well as glucose to supply carbon and energy. In practice, additional growth factors, such as trace elements and vitamins, must be added to M9 before it will support bacterial growth. Precisely what supplements are needed depends on the species concerned.

The second medium described in Table 3.1 is rather different. LB is a complex or **undefined medium**, meaning that the precise identity and quantity of its components are not known. This is because two of the ingredients, tryptone and yeast extract, are complicated mixtures of unknown chemical compounds. Tryptone in fact supplies amino acids and small peptides, while yeast extract (a dried preparation of partially digested

Table 3.1 The composition of two typical media for the growth of bacterial cultures

Component	g/l
1. M9 medium	
Na_2HPO_4	6.0
KH_2PO_4	3.0
NaCl	0.5
NH_4Cl	1.0
$MgSO_4$	0.5
glucose	2.0
$CaCl_2$	0.01
2. LB (Luria–Bertani) medium	
tryptone	10
yeast extract	5
NaCl	10

yeast cells) provides the nitrogen requirements, along with sugars and inorganic and organic nutrients. Complex media such as LB need no further supplementation and support the growth of a wide range of bacterial species.

Defined media must be used when the bacterial culture has to be grown under precisely controlled conditions. However, this is not necessary when the culture is being grown simply as a source of DNA, and under these circumstances a complex medium is appropriate. In LB medium at 37°C, aerated by shaking at 150–250 rev. min on a rotary platform, *E. coli* cells will divide once every 20 minutes or so until the culture reaches a maximum density of about $2–3 \times 10^9$ cells/ml. The growth of the culture can be monitored by reading the optical density (OD) at 600 nm (Figure 3.2), at which wavelength one OD unit corresponds to about 0.8×10^9 cells/ml.

In order to prepare a cell extract, the bacteria must be obtained in as small a volume as possible. Harvesting is therefore performed by spinning the culture in a centrifuge (Figure 3.3). Fairly low centrifugation speeds will pellet the bacteria at the bottom of the centrifuge tube, allowing the culture medium to be poured off. Bacteria from a 1000 ml culture at maximum cell density can then be resuspended into a volume of 10 ml or less.

3.1.2 PREPARATION OF A CELL EXTRACT

The bacterial cell is enclosed in a cytoplasmic membrane and surrounded by a rigid cell wall. With some species, including *E. coli*, the cell wall may itself be enveloped by a second, outer membrane. All of these barriers have to be disrupted to release the cell components.

Techniques for breaking open bacterial cells can be divided into physical methods, in which the cells are disrupted by mechanical forces, and

(a) Measurement of optical density

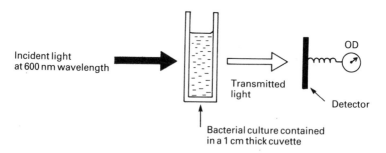

Incident light
at 600 nm wavelength

Transmitted
light

OD

Detector

Bacterial culture contained
in a 1 cm thick cuvette

(b) Estimation of cell number from a calibration curve

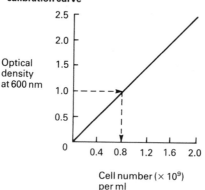

Optical density at 600 nm

Cell number ($\times 10^9$) per ml

Figure 3.2 Estimation of bacterial cell number by measurement of optical density. (a) A sample of the culture is placed in a glass cuvette and light with a wavelength of 600 nm shone through. The amount of light that passes through the culture is measured and the optical density (also called the absorbance) calculated as 1 OD unit $= -\log_{10}$ (intensity of transmitted light)/(intensity of incident light). The operation is performed with a spectrophotometer. (b) The cell number corresponding to the OD reading is calculated from a calibration curve. This curve is plotted from the OD values of a series of cultures of known cell density. For *E. coli* 1 OD unit $\approx 0.8 \times 10^9$ cells/ml.

chemical methods, where cell lysis is brought about by exposure to chemical agents that affect the integrity of the cell barriers. Chemical methods are most commonly used with bacterial cells when the object is DNA preparation.

Chemical lysis generally involves one agent attacking the cell wall and another disrupting the cell membrane (Figure 3.4a). The chemicals that are used depend on the species of bacterium involved, but with *E. coli* and related organisms, weakening of the cell wall is usually brought about by **lysozyme**, ethylenediamine tetraacetate (EDTA), or a combination of both. Lysozyme is an enzyme that is present in egg white and in secretions such as

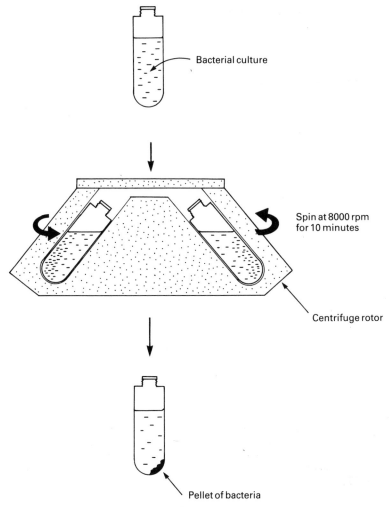

Figure 3.3 Harvesting bacteria by centrifugation.

tears and saliva, and which digests the polymeric compounds that give the cell wall its rigidity. EDTA, on the other hand, removes magnesium ions that are essential for preserving the overall structure of the cell envelope, and also inhibits cellular enzymes that could degrade DNA. Under some conditions, weakening the cell wall with lysozyme or EDTA is sufficient to cause bacterial cells to burst, but usually a detergent such as sodium dodecyl sulphate (SDS) is also added. Detergents aid the process of lysis by removing lipid molecules and thereby cause disruption of the cell membranes.

Having lysed the cells, the final step in preparation of a cell extract is removal of insoluble cell debris. Components such as partially digested cell

(a) Cell lysis

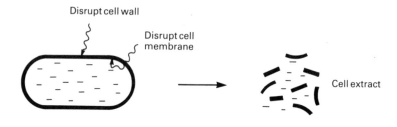

(b) Centrifugation to remove cell debris

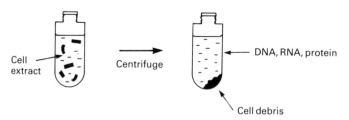

Figure 3.4 Preparation of a cell extract. (a) Cell lysis. (b) Centrifugation of the cell extract to remove insoluble debris.

wall fractions can be pelleted by centrifugation (Figure 3.4b), leaving the cell extract as a reasonably clear supernatant.

3.1.3 PURIFICATION OF DNA FROM A CELL EXTRACT

In addition to DNA, a bacterial cell extract will contain significant quantities of protein and RNA. A variety of procedures can be used to remove these contaminants, leaving the DNA in a pure form.

The standard way to deproteinize a cell extract is to add phenol or a 1:1 mixture of phenol and chloroform. These organic solvents precipitate proteins but leave the nucleic acids (DNA and RNA) in aqueous solution. The result is that if the cell extract is mixed gently with the solvent, and the layers then separated by centrifugation, precipitated protein molecules are left as a white coagulated mass at the interface between the aqueous and organic layers (Figure 3.5). The aqueous solution of nucleic acids can then be removed with a pipette.

With some cell extracts the protein content is so great that a single phenol extraction is not sufficient to purify completely the nucleic acids. This problem could be solved by carrying out several phenol extractions one after

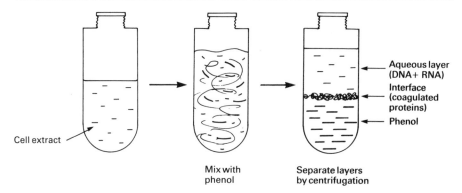

Cell extract

Mix with
phenol

Separate layers
by centrifugation

Aqueous layer
(DNA+ RNA)

Interface
(coagulated
proteins)

Phenol

Figure 3.5 Removal of protein contaminants by phenol extraction.

the other, but this is undesirable as each mixing and centrifugation step results in a certain amount of breakage of the DNA molecules. The answer is in fact to treat the cell extract with a **protease** such as Pronase or Proteinase K before phenol extraction. These enzymes will break polypeptides down into smaller units which are more easily removed by phenol.

Some RNA molecules, especially messenger RNA, will be removed by phenol treatment, but most will be retained with the DNA in the aqueous layer. The only effective way to remove the RNA is with the enzyme **ribonuclease**, which will rapidly degrade these molecules into ribonucleotide subunits.

3.1.4 CONCENTRATION OF DNA SAMPLES

Often a successful preparation will result in a very thick solution of DNA that will not need to be concentrated any further. However, dilute solutions will sometimes be obtained and it is important to consider methods for increasing the DNA concentration.

The most frequently used method of concentration is **ethanol precipitation**. In the presence of salt (strictly speaking, monovalent cations such as Na^+), and at a temperature of $-20°C$ or less, absolute ethanol will efficiently precipitate polymeric nucleic acids. With a thick solution of DNA the ethanol can be layered on top of the sample, causing molecules to precipitate at the interface. A spectacular trick is to push a glass rod through the ethanol into the DNA solution. When the rod is removed, DNA molecules will adhere and be pulled out of the solution in the form of a long fibre (Figure 3.6a). Alternatively, if ethanol is mixed with a dilute DNA solution, the precipitate can be collected by centrifugation (Figure 3.6b), and then redissolved in an appropriate volume of water. Ethanol precipitation has the added advantage of leaving in solution short-chain and monomeric nucleic acid components. Ribonucleotides produced by ribonuclease treatment will therefore be lost at this stage.

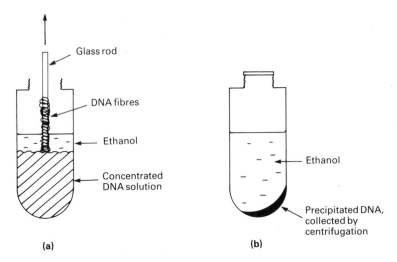

Figure 3.6 Collecting DNA by ethanol precipitation. (a) Absolute ethanol is layered on top of a concentrated solution of DNA. Fibres of DNA can be withdrawn with a glass rod. (b) For less concentrated solutions ethanol is added (at a ratio of 2.5 volumes of absolute ethanol to 1 volume of DNA solution) and precipitated DNA collected by centrifugation.

3.1.5 MEASUREMENT OF DNA CONCENTRATION

It is crucial to know exactly how much DNA is present in a solution when carrying out a gene cloning experiment. Fortunately, DNA concentrations can be accurately measured by **ultraviolet absorbance spectrophotometry**. The amount of ultraviolet radiation absorbed by a solution of DNA is directly proportional to the amount of DNA in the sample. Usually absorbance is measured at 260 nm, at which wavelength an absorbance (A_{260}) of 1.0 corresponds to 50 µg of double-stranded DNA per ml.

Ultraviolet absorbance can also be used to check the purity of a DNA preparation. With a pure sample of DNA the ratio of the absorbances at 260 nm and 280 nm (A_{260}/A_{280}) is 1.8. Ratios of less than 1.8 indicate that the preparation is contaminated, either with protein or with phenol.

3.1.6 PREPARATION OF TOTAL CELL DNA FROM ORGANISMS OTHER THAN BACTERIA

Bacteria are not the only organisms from which DNA may be required. Total cell DNA from, for example, plants or animals will be required if the aim of the genetic engineering project is to clone genes from these organisms. Although the basic steps in DNA purification are the same whatever the organism, some modifications may have to be introduced to take account of the special features of the cells being used.

Obviously growth of cells in liquid medium may not always be appropriate, even though plant and animal cell cultures are becoming increasingly important in biology. The major modifications, however, are likely to be needed at the cell breakage stage. The chemicals used for disrupting bacterial cells will not usually work with other organisms; lysozyme, for example, has no effect on plant cells. Specific degradative enzymes are available for most cell wall types, but often physical techniques, such as grinding frozen material with a mortar and pestle, will be more efficient. On the other hand, most animal cells have no cell wall at all, and can be lysed simply by treating with detergent.

3.2 PREPARATION OF PLASMID DNA

Purification of plasmids from a culture of bacteria involves the same general strategy as preparation of total cell DNA. A culture of cells, containing plasmids, will be grown in liquid medium, harvested, and a cell extract prepared. The extract will be deproteinized, the RNA removed, and the DNA probably concentrated by ethanol precipitation. However, there is an important distinction between plasmid purification and preparation of total cell DNA: in a plasmid preparation it will always be necessary to separate the plasmid DNA from the large amount of bacterial chromosomal DNA that is also present in the cells.

Separating the two types of DNA can be very difficult, but is nonetheless essential if the plasmids are to be used as cloning vehicles. The presence of the smallest amount of contaminating bacterial DNA in a gene cloning experiment may easily lead to undesirable results. Fortunately, several methods are available for removal of bacterial DNA during plasmid purification, and the use of these methods, individually or in combination, can result in isolation of very pure plasmid DNA.

These methods are based on the several physical differences between plasmid DNA and bacterial DNA, the most obvious of which is size. The largest plasmids are only 8% of the size of the E. coli chromosome, and most are much smaller than this. Techniques that can separate small DNA molecules from large ones should therefore effectively purify plasmid DNA.

In addition to size, plasmids and bacterial DNA differ in **conformation**. When applied to a polymer such as DNA, the term conformation refers to the overall spatial configuration of the molecule, with the two simplest conformations being linear and circular. Plasmids and the bacterial chromosome are circular, but during preparation of the cell extract the chromosome will always be broken to give linear fragments. A method for separating circular from linear molecules will therefore result in pure plasmids.

3.2.1 SEPARATION ON THE BASIS OF SIZE

The usual stage at which to attempt size fractionation is during preparation of the cell extract. If the cells are lysed under very carefully controlled

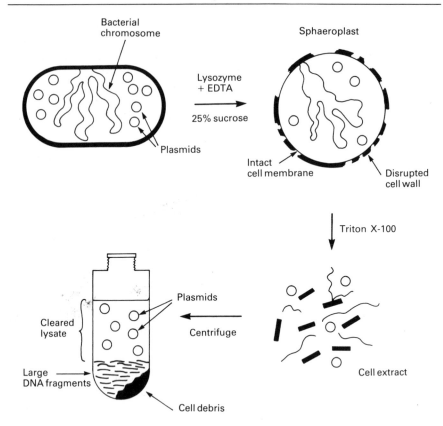

Figure 3.7 Preparation of a cleared lysate.

conditions, then only a minimal amount of chromosomal DNA breakage
will occur. The resulting DNA fragments will still be very large, much larger
than the plasmids, and will be removed with the cell debris by centrifugation. This process is aided by the fact that the bacterial chromosome is
physically attached to the cell envelope, and fragments will almost certainly
sediment with the cell debris if these attachments are not broken.

Cell disruption must therefore be performed very gently to prevent
wholesale breakage of the bacterial DNA. For *E. coli* and related species,
controlled lysis is performed as shown in Figure 3.7. Treatment with EDTA
and lysozyme is carried out in the presence of sucrose, which prevents the
cells from bursting straight away. Instead, **sphaeroplasts** are formed,
partially wall-less cells that retain an intact cytoplasmic membrane. Cell lysis
is now induced by adding a non-ionic detergent such as Triton X-100 (ionic
detergents, such as SDS, cause chromosomal breakage). This method
causes very little breakage of the bacterial DNA, so centrifugation will now
leave a **cleared lysate**, consisting almost entirely of plasmid DNA.

A cleared lysate will, however, invariably retain some chromosomal DNA. Furthermore, if the plasmids themselves are large molecules, then they may also sediment with the cell debris. Size fractionation is therefore rarely sufficient on its own, and we must consider alternative ways of removing the bacterial DNA contaminants.

3.2.2 SEPARATION ON THE BASIS OF CONFORMATION

Before considering the ways in which conformational differences between plasmids and bacterial DNA can be used to separate the two types of DNA, we must look more closely at the overall structure of plasmid DNA. It is not strictly correct to say that plasmids have a circular conformation, because double-stranded DNA circles can in fact take up one of two quite distinct configurations. Most plasmids exist in the cell as **supercoiled** molecules (Figure 3.8a). Supercoiling occurs because the double helix of the plasmid DNA is partially unwound during the plasmid replication process by enzymes called topoisomerases (p. 56). The supercoiled conformation can be maintained only if both polynucleotide strands are intact, hence the more technical name of **covalently closed-circular (ccc) DNA**. If one of the polynucleotide strands is broken, then the double helix will revert to its normal, **relaxed** state, and the plasmid will take on the alternative conformation, called **open-circular (oc)** (Figure 3.8b).

Supercoiling is important in plasmid preparation because supercoiled molecules can be fairly easily separated from non-supercoiled DNA. Two different methods are commonly used. Both can purify plasmid DNA from crude cell extracts, though in practice best results are obtained if a cleared lysate is first prepared.

(a) Alkaline denaturation

The basis to this technique is that there is a narrow pH range at which non-supercoiled DNA is **denatured**, whereas supercoiled plasmids are not. If sodium hydroxide is added to a cell extract or cleared lysate, so that the pH

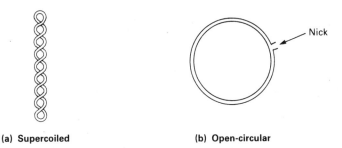

    (a) **Supercoiled**   (b) **Open-circular**

Figure 3.8 Two conformations of circular double-stranded DNA. (a) Supercoiled: both strands are intact. (b) Open-circular: one or both strands are nicked.

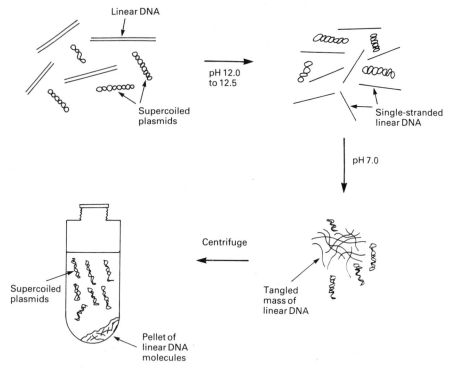

Figure 3.9 Plasmid purification by the alkaline denaturation method.

is adjusted to 12.0–12.5, then the hydrogen bonding in non-supercoiled DNA molecules will be broken, causing the double helix to unwind and the two polynucleotide chains to separate (Figure 3.9). If acid is now added, these denatured bacterial DNA strands will reaggregate into a tangled mass. Centrifugation will pellet this insoluble network, leaving pure plasmid DNA in the supernatant. An additional advantage of this procedure is that, under some circumstances (specifically, cell lysis by SDS and neutralization with sodium acetate), most of the protein and RNA will also become insoluble and be removed by the centrifugation step. Phenol extraction and ribonuclease treatment may therefore not be needed if the alkaline denaturation method is used.

(b) Ethidium bromide – caesium chloride density gradient centrifugation

This is a specialized version of the more general technique of equilibrium or **density gradient centrifugation**. A density gradient is produced by centrifuging a solution of caesium chloride (CsCl) at a very high speed (Figure 3.10a). The gradient develops because a high centrifugal force will pull the

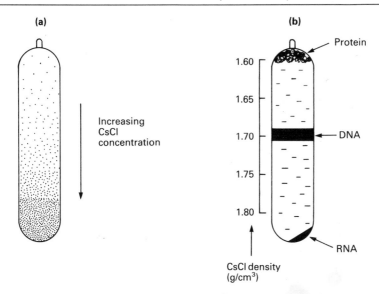

Figure 3.10 CsCl density gradient centrifugation. (a) A CsCl density gradient produced by high-speed centrifugation. (b) Separation of protein, DNA and RNA in a density gradient.

caesium and chloride ions towards the bottom of the tube. Their downward migration will be counterbalanced by diffusion, so a concentration gradient can be set up, with the CsCl density greater towards the bottom of the tube.

Macromolecules present in the CsCl solution when it is centrifuged will form bands at distinct points in the gradient (Figure 3.10b). Exactly where a particular molecule bands depends on its **buoyant density**. DNA has a buoyant density of about 1.7 g/cm^3, and will therefore migrate to the point in the gradient where the CsCl density is also 1.7 g/cm^3. In contrast, protein molecules have much lower buoyant densities, and will float at the top of the tube, whereas RNA will pellet at the bottom (Figure 3.10b). Density gradient centrifugation can therefore separate DNA, RNA and protein and is an alternative to phenol extraction and ribonuclease treatment for DNA purification.

More importantly, density gradient centrifugation in the presence of **ethidium bromide (EtBr)** can be used to separate supercoiled DNA from non-supercoiled molecules. EtBr binds to DNA molecules by intercalating between adjacent base-pairs, causing partial unwinding of the double helix (Figure 3.11). This unwinding results in a decrease in the buoyant density, by as much as 0.125 g/cm^3 for linear DNA. However, supercoiled DNA, with no free ends, has very little freedom to unwind, and can only bind a limited amount of EtBr. The decrease in buoyant density of a supercoiled molecule is therefore much less, only about 0.085 g/cm^3. As a consequence, supercoiled molecules will band in an EtBr–CsCl gradient at a different position to linear and open-circular DNA (Figure 3.12a).

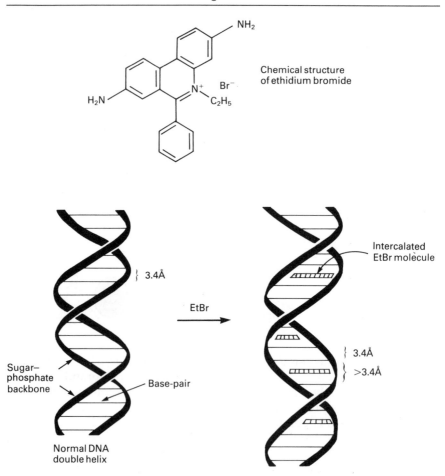

Figure 3.11 Partial unwinding of the DNA double helix by EtBr intercalation between adjacent base pairs. The normal DNA molecule shown on the left is partially unwound by taking up four EtBr molecules, resulting in the 'stretched' structure on the right.

EtBr–CsCl density gradient centrifugation is a very efficient method for obtaining pure plasmid DNA. When a cleared lysate is subjected to this procedure, plasmids band at a distinct point, separated from the linear bacterial DNA, with the protein floating on the top of the gradient and RNA pelleted at the bottom. The position of the DNA bands can be seen by shining an ultraviolet light on the tube, which causes the bound EtBr to fluoresce. The pure plasmid DNA is removed by puncturing the side of the tube and withdrawing a sample with a syringe (Figure 3.12b). The EtBr bound to the plasmid DNA is extracted with *n*-butanol (Figure 3.12c) and the CsCl removed by dialysis (Figure 3.12d). The resulting plasmid preparation will be virtually 100% pure and ready for use as a cloning vehicle.

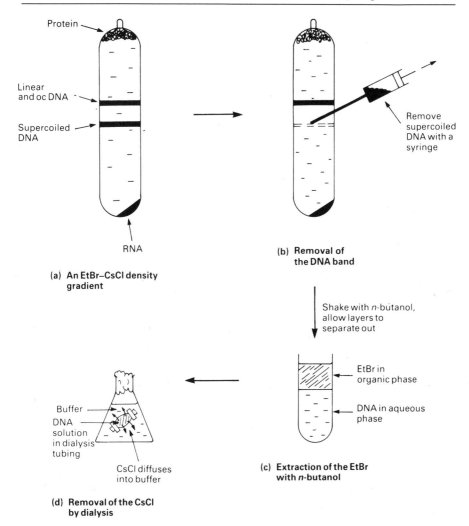

Figure 3.12 Purification of plasmid DNA by EtBr–CsCl density gradient centrifugation.

3.2.3 PLASMID AMPLIFICATION

Preparation of plasmid DNA can be hindered by the fact that plasmids make up only a small proportion of the total DNA in the bacterial cell. The yield of DNA from a bacterial culture may therefore be disappointingly low. **Plasmid amplification** offers a means of increasing this yield.

The aim of amplification is to increase the copy number of a plasmid. Some **multicopy plasmids** (those with copy numbers of 20 or more) have the useful property of being able to replicate in the absence of protein synthesis.

This contrasts with the main bacterial chromosome, which cannot replicate under these conditions. This property can be made use of during the growth of a bacterial culture for plasmid DNA purification. After a satisfactory cell density has been reached, an inhibitor of protein synthesis (for example, chloramphenicol) is added, and the culture incubated for a further 12 hours. During this time the plasmid molecules continue to replicate, even though chromosomal replication and cell division are blocked (Figure 3.13). The result is that plasmid copy numbers of several thousand may be attained. Amplification is therefore a very efficient way of increasing the yield of multicopy plasmids.

3.3 PREPARATION OF BACTERIOPHAGE DNA

The key difference between phage DNA purification and the preparation of either total cell DNA or plasmid DNA, is that for phages the starting material is not normally a cell extract. This is because bacteriophage particles can be obtained in large numbers from the extracellular medium of an infected bacterial culture. When such a culture is centrifuged, the bacteria are pelleted, leaving the phage particles in suspension (Figure 3.14). The phage particles are then collected from the suspension and their DNA extracted by a single deproteinization step to remove the phage capsid.

This overall process is rather more straightforward than the procedure used to prepare total cell or plasmid DNA. Nevertheless, successful purification of significant quantities of phage DNA is subject to several pitfalls. The main difficulty, especially with λ, is growing an infected culture in such a way that the extracellular phage titre (meaning the number of phage particles per ml of culture) is sufficiently high. In practical terms, the maximum titre that can reasonably be expected for λ is 10^{10} per ml; yet 10^{10} λ particles will yield only 500 ng of DNA. Large culture volumes, in the range of 1000 to 2000 ml are therefore needed if substantial quantities of λ DNA are to be obtained.

3.3.1 GROWTH OF CULTURES TO OBTAIN A HIGH λ TITRE

Growing a large-volume culture is no problem (bacterial cultures of 50 litres and over are common in biotechnology), but obtaining the maximum phage titre requires a certain amount of skill. The naturally occurring λ phage is lysogenic (p. 18), and an infected culture will consist mainly of cells carrying the prophage integrated into the bacterial DNA (Figure 2.7). The extracellular λ titre will be extremely low under these circumstances.

To get a high yield of extracellular λ, the culture must be **induced**, so that all the bacteria enter the lytic phase of the infection cycle, resulting in cell death and release of λ particles into the medium. Induction is normally very difficult to control, but most laboratory strains of λ carry a **temperature-sensitive (ts) mutation** in the cI gene. This is one of the genes that are

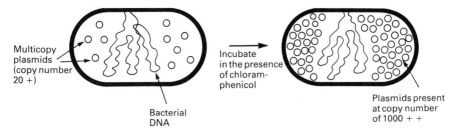

Figure 3.13 Plasmid amplification.

responsible for maintaining the phage in the integrated state. If inactivated by a mutation, the *cI* gene will no longer function correctly and the switch to lysis will occur. In the *cIts* mutation, the *cI* gene is functional at 30°C, at which temperature normal lysogeny can occur. But at 42°C, the *cIts* gene product does not work properly, and lysogeny cannot be maintained. A culture of *E. coli* infected with λ *cIts* can therefore be induced to produce extracellular phage by transferring from 30°C to 42°C (Figure 3.15).

3.3.2 PREPARATION OF NON-LYSOGENIC λ PHAGE

Although most λ strains are lysogenic, many cloning vectors derived from λ are modified, by deletions of the *cI* and other genes, so that lysogeny will

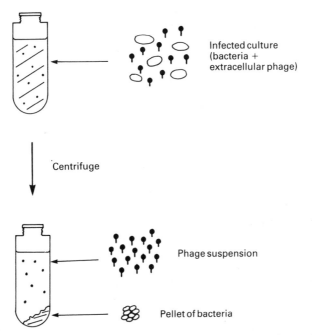

Figure 3.14 Preparation of a phage suspension from an infected culture of bacteria.

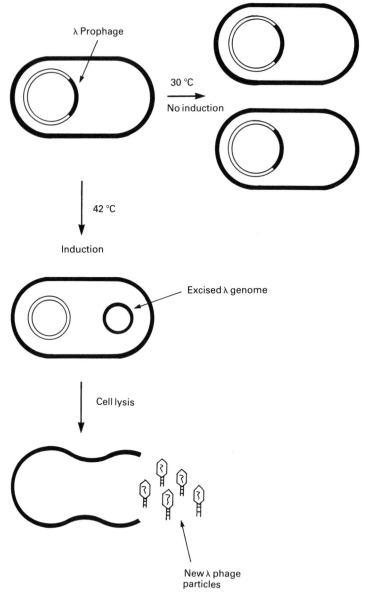

Figure 3.15 Induction of a λ *cIts* lysogen by transferring from 30°C to 42°C.

never occur. These phages cannot integrate into the bacterial genome and can infect cells only by a lytic cycle (p. 18).

With these phages the key to obtaining a high titre lies in the way in which the culture is grown, in particular the stage at which the cells are infected by

(a) Culture density is too low

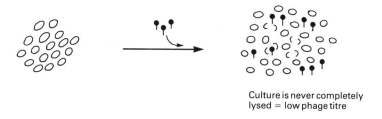

(b) Culture density is too high

(c) Culture density is just right

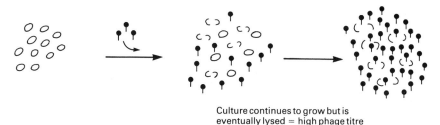

Figure 3.16 Achieving the right balance between culture age and inoculum size when preparing a sample of a non-lysogenic phage.

adding phage particles. If phages are added before the cells are dividing at their maximal rate, then all the cells will be lysed very quickly, resulting in a low titre (Figure 3.16a). On the other hand, if the cell density is too high when the phages are added, then the culture will never be completely lysed, and again the phage titre will be low (Figure 3.16b). The ideal situation is when the age of the culture, and the size of the phage inoculum, are balanced such that the culture continues to grow, but eventually all the cells are infected and lysed (Figure 3.16c). As can be imagined, skill and experience are needed to judge the matter to perfection.

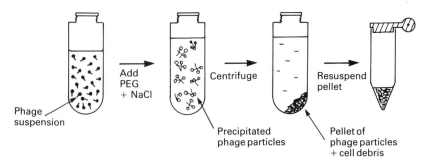

Figure 3.17 Collection of phage particles by polyethylene glycol (PEG) precipitation.

3.3.3 COLLECTION OF PHAGE FROM AN INFECTED CULTURE

The remains of lysed bacterial cells, along with any intact cells that are inadvertently left over, can be removed from an infected culture by centrifugation, leaving the phage particles in suspension (Figure 3.14). The problem now is to reduce the size of the suspension from 1000–2000 ml (the size of the original culture) to 5 ml or less, a manageable size for DNA extraction.

Phage particles are so small that they are pelleted only by very high speed centrifugation. Collection of phages is therefore usually achieved by precipitation with **polyethylene glycol (PEG)**. This is a long-chain polymeric compound which, in the presence of salt, absorbs water, thereby causing macromolecular assemblies such as phage particles to precipitate. The precipitate can then be collected by centrifugation, and redissolved in a suitably small volume (Figure 3.17).

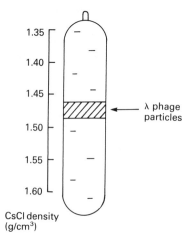

Figure 3.18 Purification of λ phage particles by CsCl density gradient centifugation.

3.3.4 PURIFICATION OF DNA FROM λ PHAGE PARTICLES

Deproteinization of the redissolved PEG precipitate is sometimes sufficient to extract pure phage DNA, but usually λ phages are subjected to an intermediate purification step. This is necessary because the PEG precipitate will also contain a certain amount of bacterial debris, possibly including unwanted cellular DNA. These contaminants can be separated from the λ particles by CsCl density gradient centrifugation. λ particles band in a CsCl gradient at 1.45 to 1.50 g/cm^3 (Figure 3.18), and can be withdrawn from the gradient just as described previously for DNA bands (p. 40 and Figure 3.12). Removal of CsCl by dialysis leaves a pure phage preparation from which the DNA can be extracted by either phenol or protease treatment to digest the phage protein coat.

3.3.5 PURIFICATION OF M13 DNA CAUSES FEW PROBLEMS

Most of the differences between the M13 and λ infection cycles are to the advantage of the molecular biologist wishing to prepare M13 DNA. First, the double-stranded replicative form of M13 (p. 24), which behaves like a high copy number plasmid, is very easily purified by following the standard procedures for plasmid preparation. A cell extract is prepared from cells infected with M13, and the replicative form separated from bacterial DNA by, for example, EtBr–CsCl density gradient centrifugation.

However, the single-stranded form of the M13 genome, contained in the extracellular phage particles, is frequently required. In this respect, the big advantage compared with λ is that high titres of M13 are very easy to obtain. As infected cells continually secrete M13 particles into the medium (Figure 2.8), with lysis never occurring, a high M13 titre is achieved simply by growing the infected culture to a high cell density. In fact titres of 10^{12} per ml and above are quite easy to obtain without any special tricks being used. Such high titres mean that significant amounts of single-stranded M13 DNA can be prepared from cultures of small volume – 5 ml or less. Furthermore, as the infected cells are not lysed, there is no problem with cell debris contaminating the phage suspension. Consequently the CsCl density gradient centrifugation step, needed for λ phage preparation, is rarely required with M13.

In summary, single-stranded M13 DNA preparation involves growth of a small volume of infected culture, centrifugation to pellet the bacteria, precipitation of the phage particles with PEG, phenol extraction to remove the phage protein coats, and ethanol precipitation to concentrate the resulting DNA (Figure 3.19).

(a) Culture of infected cells

(b) Centrifuge to remove cells

(c) Add PEG to phage suspension, centrifuge

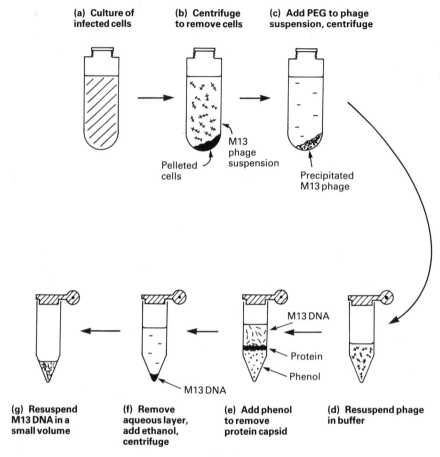

Pelleted cells

M13 phage suspension

Precipitated M13 phage

M13 DNA

Protein

Phenol

M13 DNA

(g) Resuspend M13 DNA in a small volume

(f) Remove aqueous layer, add ethanol, centrifuge

(e) Add phenol to remove protein capsid

(d) Resuspend phage in buffer

Figure 3.19 Preparation of M13 DNA from an infected culture of bacteria.

FURTHER READING

Marmur, J. (1961) A procedure for the isolation of deoxyribonucleic acid from micro-organisms. *Journal of Molecular Biology*, **3**, 208–18 – total cell DNA preparation.

Birnboim, H. C. and Doly, J. (1979) A rapid alkaline extraction procedure for screening recombinant plasmid DNA. *Nucleic Acids Research*, **7**, 1513–23 – a method for preparing plasmid DNA.

Clewell, D. B. (1972) Nature of ColE1 plasmid replication in *Escherichia coli* in the presence of chloramphenicol. *Journal of Bacteriology*, **110**, 667–76 – the biological basis to plasmid amplification.

Zinder, N. D. and Boeke, J. D. (1982) The filamentous phage (Ff) as vectors for recombinant DNA. *Gene*, **19**, 1–10 – methods for phage growth and DNA preparation.

Manipulation of purified DNA

4

Once pure samples of DNA have been prepared, the next step in a gene cloning experiment is construction of the recombinant DNA molecule (Figure 1.1). To produce this recombinant molecule, the vector, as well as the DNA to be cloned, must be cut at specific points and then joined together in a controlled manner. Cutting and joining are two examples of DNA manipulative techniques, a wide variety of which have been developed over the last few years. As well as being cut and joined, DNA molecules can be shortened, lengthened, copied into RNA or into new DNA molecules, and modified by the addition or removal of specific chemical groups. These manipulations, all of which can be carried out in the test-tube, provide the foundation not only for gene cloning, but also for basic studies into DNA biochemistry, gene structure, and the control of gene expression.

Almost all DNA manipulative techniques make use of purified enzymes. Within the cell these enzymes participate in essential processes such as DNA replication and transcription, breakdown of unwanted or foreign DNA (for example, invading virus DNA), repair of mutated DNA, and **recombination** between different DNA molecules. After purification from cell extracts, many of these enzymes can be persuaded to carry out their natural reactions, or something closely related to them, under artificial conditions. Although these enzymatic reactions are often quite straightforward, most are absolutely impossible to perform by standard chemical methods. Purified enzymes are therefore crucial to genetic engineering and an important industry has sprung up around their preparation, characterization and marketing. Commercial suppliers of high-purity enzymes provide an essential service to the molecular biologist.

The cutting and joining manipulations that underlie gene cloning are carried out by enzymes called restriction endonucleases (for cutting) and ligases (for joining). Most of this chapter will be concerned with the ways in which these two types of enzyme are used. Firstly, however, we must consider the whole range of DNA manipulative enzymes, to see exactly what types of reaction can be performed. Many of these enzymes will be mentioned in later chapters when procedures that make use of them will be described.

4.1 THE RANGE OF DNA MANIPULATIVE ENZYMES

These enzymes can be grouped into five broad classes depending on the type of reaction that they catalyse.

1. **Nucleases** are enzymes that cut, shorten or degrade nucleic acid molecules.
2. **Ligases** join nucleic acid molecules together.
3. **Polymerases** make copies of molecules.
4. **Modifying enzymes** remove or add chemical groups.
5. **Topoisomerases** introduce or remove supercoils from covalently closed-circular DNA.

Before considering in detail each of these classes of enzyme, two points should be made. The first is that, although most enzymes can be assigned to a particular class, a few display multiple activities that span two or more classes. Most importantly, many polymerases combine their ability to make new DNA molecules with an associated DNA degradative (= nuclease) activity.

Secondly it should be appreciated that, as well as the DNA manipulative enzymes, many similar enzymes able to act on RNA are known. The ribonuclease used to remove contaminating RNA from DNA preparations (p. 33) is an example of such an enzyme. Although some RNA manipulative enzymes have applications in gene cloning and will be mentioned in later chapters, we will in general restrict our thoughts to those enzymes that act on DNA.

4.1.1 NUCLEASES

Nucleases degrade DNA molecules by breaking the phosphodiester bonds that link one nucleotide to the next in a DNA strand. There are two distinct kinds of nuclease (Figure 4.1).

1. **Exonucleases**, which remove nucleotides one at a time from the end of a DNA molecule.
2. **Endonucleases**, which are able to break internal phosphodiester bonds within a DNA molecule.

The main distinction between different exonucleases lies in the number of strands that are degraded when a double-stranded molecule is attacked. The enzyme called Bal31 (purified from the bacterium *Alteromonas espejiana*) is an example of an exonuclease that removes nucleotides from both strands of a double-stranded molecule (Figure 4.2a). The greater the length of time that Bal31 is allowed to act on a group of DNA molecules, the shorter the resulting DNA fragments will be. In contrast, enzymes such as *E. coli* exonuclease III degrade just one strand of a double-stranded molecule, leaving single-stranded DNA as the product (Figure 4.2b).

The same criterion can be used to classify endonucleases. S1 endonu-

(a) An exonuclease

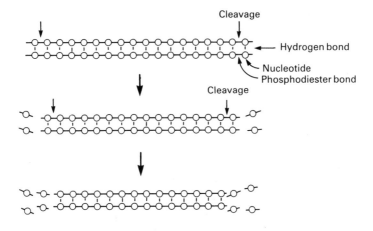

(b) An endonuclease

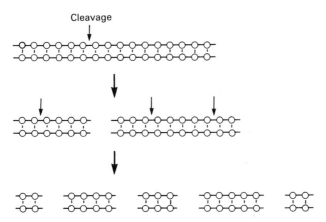

Figure 4.1 The reactions catalysed by the two different kinds of nuclease. (a) An exonuclease, which removes nucleotides from the end of a DNA molecule. (b) An endonuclease, which breaks internal phosphodiester bonds.

clease (from the fungus *Aspergillus oryzae*) will cleave only single strands (Figure 4.3a), whereas deoxyribonuclease I (DNase I), which is prepared from cow pancreas, cuts both single- and double-stranded molecules (Figure 4.3b). DNase I is non-specific in that it attacks DNA at any internal phosphodiester bond; the end result of prolonged DNase I action is therefore a mixture of mononucleotides and very short oligonucleotides. On

(a) Bal 31

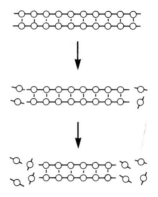

(b) Exonuclease III

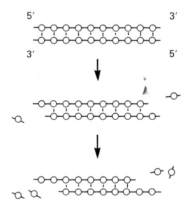

Figure 4.2 The reactions catalysed by different types of exonuclease. (a) Bal31, which removes nucleotides from both strands of a double-stranded molecule. (b) Exonuclease III, which removes nucleotides only from the 3'-terminus (see p. 78 for a description of the differences between the 3'-and 5'-termini of a polynucleotide).

the other hand, the special group of enzymes called restriction endonucleases cleave double-stranded DNA only at a limited number of specific recognition sites (Figure 4.3c). These important enzymes are described in detail on p. 56.

(a) S1 nuclease

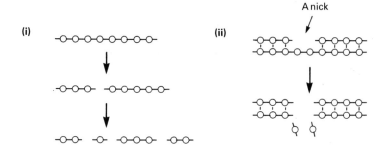

(b) DNase I

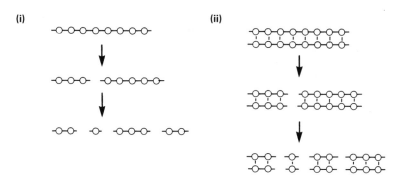

(c) A restriction endonuclease

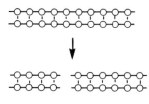

No further cleavage

Figure 4.3 The reactions catalysed by different types of endonuclease. (a) S1 nuclease, which cleaves only single-stranded DNA, including single-stranded nicks in mainly double-stranded molecules. (b) DNase I, which cleaves both single- and double-stranded DNA. (c) A restriction endonuclease, which cleaves double-stranded DNA, but only at a limited number of sites.

4.1.2 LIGASES

In the cell the function of DNA ligase is to repair single-stranded breaks ('discontinuities') that arise in double-stranded DNA molecules during, for example, DNA replication. DNA ligases from most organisms will also join together two individual fragments of double-stranded DNA (Figure 4.4) The role of these enzymes in construction of recombinant DNA molecules is described on p. 73.

4.1.3 POLYMERASES

DNA polymerases are enzymes that synthesize a new strand of DNA complementary to an existing DNA or RNA **template** (Figure 4.5a). Most polymerases can function only if the template possesses a double-stranded region which acts as a **primer** for initiation of polymerization.

Three types of DNA polymerase are used routinely in genetic engineering. The first is DNA polymerase I, which is prepared usually from *E. coli*. This enzyme attaches to a short single-stranded region (or 'nick') in a mainly double-stranded DNA molecule, and then synthesizes a completely new strand, degrading the existing strand as it proceeds (Figure 4.5b). DNA polymerase I is therefore an example of an enzyme with a dual activity – DNA polymerization and DNA degradation.

(a) Discontinuity repair

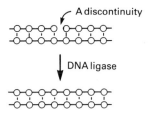

(b) Joining two molecules

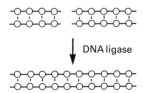

Figure 4.4 The two reactions catalysed by DNA ligase. (a) Repair of a discontinuity – a missing phosphodiester bond in one strand of a double-stranded molecule. (b) Joining two molecules together.

(a) The basic reaction

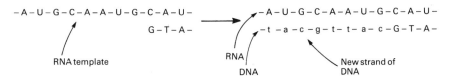

(b) DNA polymerase I

```
 - A - T - G - C - A - A - T - G - C - A - T -          - A - T - G - C - A - A - T - G - C - A - T -
 - T - A - C - G            G - T - A -                  - t - a - c - g - t - t - a - c   G - T - A -
                   _____/
                    A nick                              Existing nucleotides
                                                        are replaced
```

(c) The Klenow fragment

```
 - A - T - G - C - A - A - T - G - C - A - T -          - A - T - G - C - A - A - T - G - C - A - T -
 - T - A - C - G            G - T - A -                  - T - A - C - G - t - t - a - c - G - T - A -

                                                        Existing          Only the nick
                                                        nucleotides are   is filled in
                                                        not replaced
```

(d) Reverse transcriptase

```
 - A - U - G - C - A - A - U - G - C - A - U -          - A - U - G - C - A - A - U - G - C - A - U -
                   G - T - A -                           - t - a - c - g - t - t - a - c - G - T - A -

      RNA template                              RNA                      New strand of
                                                DNA                      DNA
```

Figure 4.5 The reactions catalysed by DNA polymerases. (a) The basic reaction: a new DNA strand is synthesized in the 5' to 3' direction. (b) DNA polymerase I, which initially fills in nicks but then continues to synthesize a new strand, degrading the existing one as it proceeds. (c) The Klenow fragment, which only fills in nicks. (d) Reverse transcriptase, which uses a template of RNA.

In fact the polymerase and nuclease activities of DNA polymerase I are controlled by different parts of the enzyme molecule, and can be separated if the enzyme is cleaved in a particular way (Figure 4.5c). The part of the enzyme that retains the polymerase function is called the **Klenow fragment**. As the Klenow fragment lacks the nuclease activity it can synthesize a complementary DNA strand only on a single-stranded template. Several other enzymes are known that perform the same function. The major application of the Klenow fragment and these related polymerases is in DNA sequencing (p. 186).

The third type of DNA polymerase that is important in genetic engineering is **reverse transcriptase**, an enzyme involved in the replication of several kinds of virus. Reverse transcriptase is unique in that it uses as a template not DNA but RNA (Figure 4.5d). The ability of this enzyme to synthesize a DNA strand complementary to an RNA template is central to the technique called cDNA cloning (p. 161).

4.1.4 DNA MODIFYING ENZYMES

There are numerous enzymes that modify DNA molecules by addition or removal of specific chemical groups. The most important are as follows.

1. **Alkaline phosphatase** (from *E. coli* or calf intestinal tissue) which removes the phosphate group present at the **5'-terminus** of a DNA molecule (Figure 4.6a).
2. **Polynucleotide kinase** (from *E. coli* infected with T4 phage) which has the reverse effect to alkaline phosphatase, adding phosphate groups on to free 5'-termini (Figure 4.6b).
3. **Terminal deoxynucleotidyl transferase** (from calf thymus tissue) which adds one or more deoxynucleotides on to the **3'-terminus** of a DNA molecule (Figure 4.6c).

4.1.5 TOPOISOMERASES

The final class of DNA manipulative enzymes are the topoisomerases, which are able to change the conformation of covalently closed-circular DNA (e.g. plasmid DNA molecules) by introducing or removing supercoils (p. 37). Although important in the study of DNA replication, topoisomerases have yet to find a real use in genetic engineering.

4.2 ENZYMES FOR CUTTING DNA – RESTRICTION ENDONUCLEASES

Gene cloning requires that DNA molecules be cut in a very precise and reproducible fashion. This is illustrated by the way in which the vector is cut during construction of a recombinant DNA molecule (Figure 4.7a). Each vector molecule must be cleaved at a single position, to open up the circle so that new DNA can be inserted; a molecule that is cut more than once will be broken into two or more separate fragments and will be of no use as a cloning vehicle. Furthermore, each vector molecule must be cut at exactly the same position on the circle – as will become apparent in later chapters, random cleavage is not satisfactory. It should be clear that a very special type of nuclease is needed to carry out this manipulation.

Often it will also be necessary to cleave the DNA that is to be cloned (Figure 4.7b). There are two reasons for this. Firstly, if the aim is to clone a

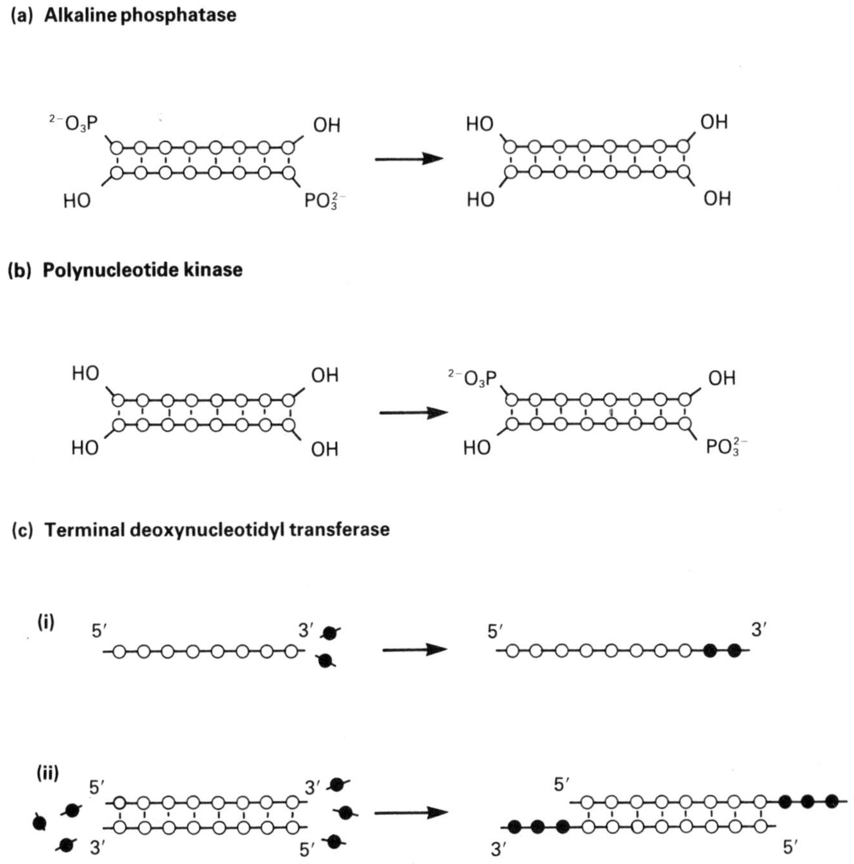

(a) Alkaline phosphatase

(b) Polynucleotide kinase

(c) Terminal deoxynucleotidyl transferase

(i)

(ii)

Figure 4.6 The reactions catalysed by DNA-modifying enzymes. (a) Alkaline phosphatase, which removes 5'-phosphate groups. (b) Polynucleotide kinase, which attaches 5'-phosphate groups. (c) Terminal deoxynucleotidyl transferase, which attaches nucleotides to the 3'-termini of polynucleotides in either (i) single-stranded or (ii) double-stranded molecules.

single gene, which may consist of only 2 or 3 kb of DNA, then that gene will have to be cut out of the large (often greater than 80 kb) DNA molecules produced by skilful use of the preparative techniques described in Chapter 3. Secondly, large DNA molecules may have to be broken down simply to produce fragments small enough to be carried by the vector. Most cloning vectors exhibit a preference for DNA fragments that fall into a particular size range; M13-based vectors, for example, are very inefficient at cloning DNA molecules of more than 3 kb in length.

Purified restriction endonucleases allow the molecular biologist to cut DNA molecules in the precise, reproducible manner required for gene cloning. The discovery of these enzymes, which led to Nobel Prizes for

(a) Vector molecules

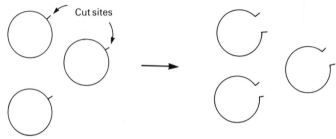

Each vector molecule must be
cut once, each at the same position

(b) The DNA molecule containing the gene to be cloned

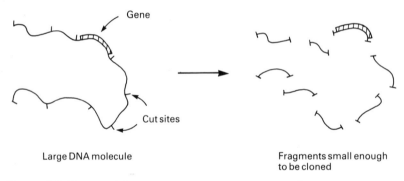

Large DNA molecule

Fragments small enough
to be cloned

Figure 4.7 The need for very precise cutting manipulations in a gene cloning experiment.

W. Arber, H. Smith and D. Nathans in 1978, was one of the key break-throughs in the development of genetic engineering.

4.2.1 THE DISCOVERY AND FUNCTION OF RESTRICTION ENDONUCLEASES

The initial observation that led to the eventual discovery of restriction endonucleases was made in the early 1950s, when it was shown that some strains of bacteria are immune to bacteriophage infection, a phenomenon referred to as **host-controlled restriction**.

The mechanism of restriction is not very complicated, even though it took

(a) Restriction of phage DNA

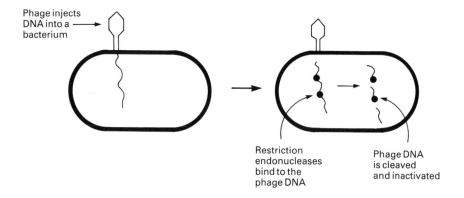

Phage injects DNA into a bacterium →

Restriction endonucleases bind to the phage DNA

Phage DNA is cleaved and inactivated

(b) Bacterial DNA is not cleaved

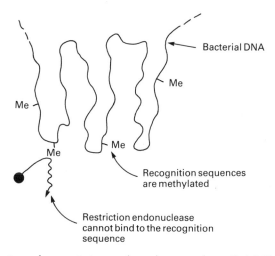

Bacterial DNA

Me

Me

Me

Me

Recognition sequences are methylated

Restriction endonuclease cannot bind to the recognition sequence

Figure 4.8 The function of a restriction endonuclease in the cell. (a) Phage DNA is cleaved, but (b) bacterial DNA is not.

over 20 years to be fully understood. Restriction occurs because the bacterium produces an enzyme that degrades the phage DNA before it has time to replicate and direct synthesis of new phage particles (Figure 4.8a). The bacterium's own DNA, the destruction of which would of course be lethal, is protected from attack because it carries additional methyl groups that block the degradative enzyme action (Figure 4.8b). These degradative enzymes are called restriction endonucleases and are synthesized by many, perhaps all, species of bacteria; over 1200 different ones have now been characterized. Three different classes of restriction endonuclease are recognized, each distinguished by a slightly different mode of action. Types I and III are rather complex and have only a very limited role in genetic engineering. Type II restriction endonucleases, on the other hand, are the cutting enzymes that are so important in gene cloning.

4.2.2 TYPE II RESTRICTION ENDONUCLEASES CUT DNA AT SPECIFIC NUCLEOTIDE SEQUENCES

The central feature of Type II restriction endonucleases (which will be referred to simply as 'restriction endonucleases' from now on) is that each enzyme has a specific recognition sequence at which it cuts a DNA molecule. A particular enzyme will cleave DNA at the recognition sequence and nowhere else. For example, the restriction endonuclease called *PvuI* (isolated from *Proteus vulgaris*) cuts DNA only at the hexanucleotide CGATCG. In contrast, a second enzyme from the same bacterium, called *PvuII*, cuts at a different hexanucleotide, in this case CAGCTG.

Many restriction endonucleases recognize hexanucleotide target sites, but others cut at four, five or even eight nucleotide sequences. *Sau3A* (from *Staphylococcus aureus* strain 3A) recognizes GATC, and *AluI* (*Arthrobacter luteus*) cuts at AGCT. There are also examples of restriction endonucleases

Table 4.1 The recognition sequences for some of the most frequently used restriction endonucleases

Enzyme	Organism	Recognition[a] sequence	Blunt or sticky end
*Eco*RI	*Escherichia coli*	GAATTC	sticky
*Bam*HI	*Bacillus amyloliquefaciens*	GGATCC	sticky
Bgl/II	*Bacillus globigii*	AGATCT	sticky
*Pvu*I	*Proteus vulgaris*	CGATCG	sticky
*Pvu*II	*Proteus vulgaris*	CAGCTG	blunt
*Hind*III	*Haemophilus influenzae* R_d	AAGCTT	sticky
*Hinf*I	*Haemophilus influenzae* R_f	GANTC	sticky
*Sau*3A	*Staphylococcus aureus*	GATC	sticky
*Alu*I	*Arthrobacter luteus*	AGCT	blunt
*Taq*I	*Thermus aquaticus*	TCGA	sticky
*Hae*III	*Haemophilus aegyptius*	GGCC	blunt
*Not*I	*Nocardia otitidis-caviarum*	GCGGCCGC	sticky

[a] The sequence shown is that of one strand, given in the 5′ to 3′ direction. Note that almost all recognition sequences are palindromes: when both strands are considered they read the same in each direction, e.g.

```
         5′ G|A  A  T  T  C 3′
EcoRI:      |  |__+__+__+__|  |
         3′ C  T  T  A  A|G 5′
```

with degenerate recognition sequences, meaning that they cut DNA at any one of a family of related sites. HinfI (*Haemophilus influenzae* strain R$_f$), for instance, recognizes GANTC, so cuts at GAATC, GATTC, GAGTC and GACTC.

The restriction sequences for some of the most frequently used restriction endonucleases are listed in Table 4.1.

4.2.3 BLUNT ENDS AND STICKY ENDS

The exact nature of the cut produced by a restriction endonuclease is of considerable importance in the design of a gene cloning experiment. Many restriction endonucleases make a simple double-stranded cut in the middle of the recognition sequence (Figure 4.9a), resulting in a **blunt** or flush end. *Pvu*II and *Alu*I are examples of blunt-end cutters.

However, quite a large number of restriction endonucleases cut DNA in a slightly different way. With these enzymes the two DNA strands are not cut at exactly the same position. Instead the cleavage is staggered, usually by two or four nucleotides, so that the resulting DNA fragments have short single-stranded overhangs at each end (Figure 4.9b). These are called sticky or cohesive ends, as base pairing between them can stick the DNA molecule back together again (recall that sticky ends were encountered on p. 22 during the description of λ phage replication). One important feature of sticky end enzymes is that restriction endonucleases with different recognition sequences may produce the same sticky ends. *Bam*HI (recognition sequence GGATCC) and *Bgl*II (AGATCT) are examples – both produce GATC sticky ends (Figure 4.9c). The same sticky end is also produced by *Sau*3A, which recognizes only the tetranucleotide GATC. Fragments of DNA produced by cleavage with either of these enzymes can be joined to each other, as each fragment will carry a complementary sticky end.

4.2.4 THE FREQUENCY OF RECOGNITION SEQUENCES IN A DNA MOLECULE

The number of recognition sequences for a particular restriction endonuclease in a DNA molecule of known length can be calculated mathematically. A tetranucleotide sequence (e.g. GATC) should occur once every $4^4 = 256$ nucleotides, and a hexanucleotide (e.g. GGATCC) once every $4^6 = 4096$ nucleotides. These calculations assume that the nucleotides are ordered in a random fashion and that the four different nucleotides are present in equal proportions (i.e. the GC content $= 50\%$). In practice, neither of these assumptions is entirely valid. For example, the λ DNA molecule, at 49 kb, should contain about 12 sites for a restriction endonuclease with a hexanucleotide recognition sequence. In fact these recognition sites occur less frequently (for example, six for *Bgl*II, five for *Bam*HI and only two for *Sal*I), a reflection of the fact that the GC content for λ is rather less than 50% (Figure 4.10a).

(a) Production of blunt ends

-N-N-A-G-C-T-N-N-   *AluI*   -N-N-A-G-   C-T-N-N-

-N-N-T-C-G-A-N-N-   ──────▶   -N-N-T-C   G-A-N-N-

'N' = A, G, C or T   Blunt ends

(b) Production of sticky ends

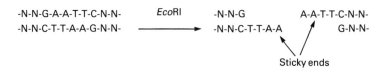

-N-N-G-A-A-T-T-C-N-N-   *EcoR*I   -N-N-G   A-A-T-T-C-N-N-

-N-N-C-T-T-A-A-G-N-N-   ──────▶   -N-N-C-T-T-A-A   G-N-N-

                                                         Sticky ends

(c) The same sticky ends produced by different restriction endonucleases

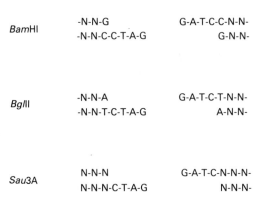

*Bam*HI   -N-N-G   G-A-T-C-C-N-N-
           -N-N-C-C-T-A-G   G-N-N-

*Bgl*II   -N-N-A   G-A-T-C-T-N-N-
           -N-N-T-C-T-A-G   A-N-N-

*Sau*3A   N-N-N   G-A-T-C-N-N-N-
           N-N-N-C-T-A-G   N-N-N-

Figure 4.9 The ends produced by cleavage of DNA with different restriction endonucleases. (a) A blunt end produced by *AluI*. (b) A sticky end produced by *EcoR*I. (c) The same sticky ends produced by *Bam*HI, *Bgl*II and *Sau*3A.

Furthermore, restriction sites are generally not evenly spaced out along a DNA molecule. If they were then digestion with a particular restriction endonuclease would give fragments of roughly equal sizes. Figure 4.10b shows the fragments produced by cutting λ DNA with *Bgl*II, *Bam*HI and *Sal*I. In each case there is a considerable spread of fragment sizes, indicating that in λ DNA the nucleotides are not randomly ordered.

The lesson to be learned from Figure 4.10 is that, although mathematics

(a) Cleavage sites on λ DNA

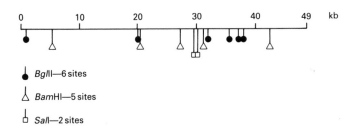

↓ *Bgl*II—6 sites

△ *Bam*HI—5 sites

⊔ *Sal*I—2 sites

(b) Fragment sizes

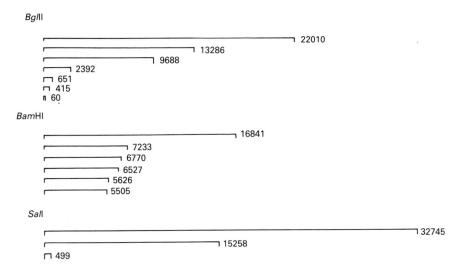

Figure 4.10 Restriction of the λ DNA molecule. (a) The positions of the recognition sequences for *Bgl*II, *Bam*HI and *Sal*I. (b) The fragments produced by cleavage with each of these restriction endonucleases; the numbers are the fragment sizes in base pairs.

may give an idea of how many restriction sites to expect in a given DNA molecule, only experimental analysis can provide the true picture. We must therefore move on to consider how restriction endonucleases are used in the laboratory.

4.2.5 PERFORMING A RESTRICTION DIGEST IN THE LABORATORY

As an example we will consider how to digest a sample of λ DNA (concentration 125 μg/ml) with *Bgl*II.

First of all the required amount of DNA must be pipetted into a test-tube. The amount of DNA that will be restricted depends on the nature of the experiment; in this case we shall digest 2 µg of λ DNA, which is contained in 16 µl of the sample (Figure 4.11a). Very accurate micropipettes will therefore be needed.

Clearly the other main component in the reaction will be the restriction endonuclease, obtained from a commercial supplier as a pure solution of known concentration. But before adding the enzyme, the solution containing the DNA must be adjusted to provide the correct conditions to ensure maximal activity of the enzyme. Most restriction endonucleases will function adequately at pH 7.4, but different enzymes vary in their requirements for ionic strength (provided usually by NaCl) and Mg^{2+} concentration (all Type II restriction endonucleases require Mg^{2+} in order to function). It is also advisable to add a reducing agent, such as dithiothreitol, which will stabilize the enzyme and prevent its inactivation. Providing the

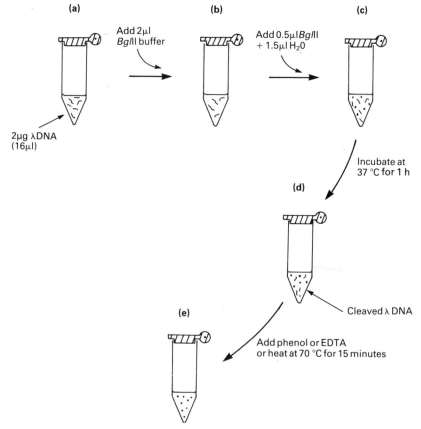

(a)

Add 2 µl BglII buffer

2µg λDNA (16µl)

(b)

Add 0.5µl BglII + 1.5µl H₂0

(c)

Incubate at 37 °C for 1 h

(d)

Cleaved λ DNA

(e)

Add phenol or EDTA or heat at 70 °C for 15 minutes

Figure 4.11 Performing a restriction digest in the laboratory. See the text for details.

Table 4.2 A 10 × buffer suitable for restriction of DNA with
*Bgl*II

Component	Concentration (mM)
Tris-HCl, pH 7.4	500
MgCl$_2$	100
NaCl	500
Dithiothreitol	10

right conditions for the enzyme is very important – incorrect NaCl or Mg^{2+} concentrations may not only decrease the activity of the restriction endonuclease, but may also cause changes in the specificity of the enzyme, so that DNA cleavage occurs at additional, non-standard recognition sequences.

The composition of a suitable buffer for *Bgl*II is shown in Table 4.2. This buffer is ten times the working concentration, and is diluted by adding it to the reaction mixture. In our example, a suitable final volume for the reaction mixture would be 20 µl, so we add 2 µl of 10 × *Bgl*II buffer to the 16 µl of DNA already present (Figure 4.11b).

The restriction endonuclease can now be added. By convention, one unit of enzyme is defined as the quantity needed to cut 1 µg of DNA in one hour, so we need two units of *Bgl*II to cut 2 µg of λ DNA. *Bgl*II is frequently obtained at a concentration of 4 units/µl, so 0.5 µl will be sufficient to cleave the DNA. The final ingredients in the reaction mixture are therefore 0.5 µl *Bgl*II + 1.5 µl water, giving a final volume of 20 µl (Figure 4.11c).

The last factor to consider is incubation temperature. Most restriction endonucleases work best at 37°C, but a few have different requirements. *Taq*I, for example, is purified from a bacterium called *Thermus aquaticus*, which normally lives at very high temperatures in habitats such as hot springs. Restriction digests with *Taq*I must be incubated at 65°C to obtain maximum enzyme activity.

After one hour the restriction should be complete (Figure 4.11d). If the DNA fragments produced by restriction are to be used in cloning experiments, then the enzyme must somehow be destroyed so that it does not accidentally digest other DNA molecules that may be added at a later stage. There are several ways of 'killing' the enzyme. For many a short incubation at 70°C is sufficient, for others phenol extraction or the addition of EDTA (which binds Mg^{2+} ions preventing restriction endonuclease action) is used (Figure 4.11e).

4.2.6 ANALYSING THE RESULT OF RESTRICTION ENDONUCLEASE CLEAVAGE

A restriction digest will result in a number of DNA fragments, the sizes of which depend on the exact positions of the recognition sequences for the endonuclease in the original molecule (Figure 4.10). Clearly a way of determining the number and sizes of the fragments is needed if restriction endonucleases are to be of use in gene cloning. Whether or not a DNA molecule is cut at all can be determined fairly easily by testing the viscosity of the solution. Larger DNA molecules result in a more viscous solution than smaller ones, so cleavage is associated with a decrease in viscosity. However, working out the number and sizes of the individual cleavage products is more difficult. In fact, for several years this was one of the most tedious aspects of experiments involving DNA. Eventually the problems were solved in the early 1970s when the technique of gel electrophoresis was developed.

(a) Separation of molecules by gel electrophoresis

DNA molecules, like proteins and many other biological compounds, carry an electric charge, negative in the case of DNA. Consequently, when DNA molecules are placed in an electric field they will migrate towards the positive pole (Figure 4.12a). The rate of migration of a molecule depends on two factors, its shape and its electric charge. Unfortunately, most DNA molecules are the same shape and all have very similar electric charges. Fragments of different sizes cannot therefore be separated by standard **electrophoresis**.

However, the size of the DNA molecule does become a factor if the electrophoresis is performed in a gel. A gel, which is usually made of agarose, polyacrylamide or a mixture of the two, comprises a complex network of pores, through which the DNA molecules must travel to reach the positive electrode. The smaller the DNA molecule, the faster it can migrate through the gel. **Gel electrophoresis** will therefore separate DNA molecules according to their size (Figure 4.12b).

In practice the composition of the gel determines the sizes of the DNA molecules that can be separated. A 0.5 cm thick slab of 0.3% agarose, which has relatively large pores, would be used for molecules in the size range 5 to 60 kb, allowing, for example, molecules of 30 and 35 kb to be clearly distinguished. At the other end of the scale, a very thin (0.3 mm) 40% polyacrylamide gel, with extremely small pores, would be used to separate much smaller DNA molecules, in the range 1 to 300 bp, and could distinguish molecules differing in length by just a single nucleotide.

(b) Visualizing DNA molecules in a gel

(i) Staining

The easiest way to see the results of a gel electrophoresis experiment is to stain the gel with a compound that will make the DNA visible. Ethidium bromide, already described on p. 40 as a means of visualizing DNA in CsCl gradients, is also routinely used to stain DNA in agarose and polyacryl-

(a) Standard electrophoresis

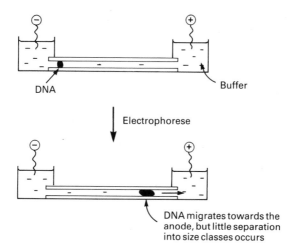

(b) Gel electrophoresis

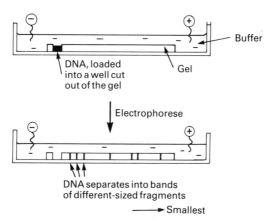

Figure 4.12 Standard electrophoresis (a) does not separate DNA fragments of different sizes, whereas gel electrophoresis (b) does.

amide gels (Figure 4.13). Bands showing the positions of the different size classes of DNA fragment are clearly visible under ultraviolet irradiation after EtBr staining, so long as sufficient DNA is present.

(ii) Autoradiography of radioactively labelled DNA

The one drawback with EtBr staining is that there is a limit to its sensitivity. If less than about 25 ng of DNA is present per band then it is unlikely that the

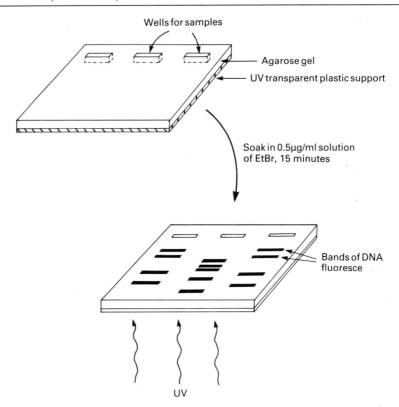

Figure 4.13 Visualizing DNA bands in an agarose gel by EtBr staining and ultraviolet irradiation.

results will show up with EtBr staining. For small amounts of DNA a much more sensitive detection method is needed.

Autoradiography provides an answer. If the DNA is **labelled**, before electrophoresis, by incorporation of a **radioactive marker** into the individual molecules, then the DNA can be visualized by placing an X-ray-sensitive photographic film over the gel. The radioactive DNA will expose the film, revealing the banding pattern (Figure 4.14).

A DNA molecule is usually labelled by incorporating nucleotides that carry the radioactive isotope of phosphorus, ^{32}P (Figure 4.15a). Several methods are available, the most popular being **nick translation** and **end-filling**.

Nick translation refers to the activity of DNA polymerase I (p. 54). Most purified samples of DNA contain some nicked molecules, however carefully the preparation has been carried out, so DNA polymerase I will be able to attach to the DNA and catalyse a strand replacement reaction (Figure 4.5b). This reaction requires a supply of nucleotides: if one of these is radioactively labelled, then the DNA molecule will itself become labelled (Figure 4.15b).

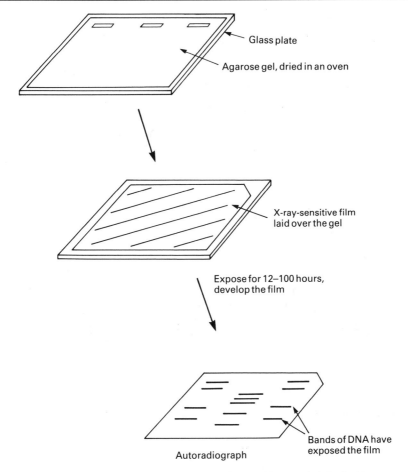

Glass plate

Agarose gel, dried in an oven

X-ray-sensitive film
laid over the gel

Expose for 12–100 hours,
develop the film

Autoradiograph

Bands of DNA have
exposed the film

Figure 4.14 The use of autoradiography to visualize radioactively labelled DNA in an agarose gel.

Nick translation can be used to label any DNA molecule but may under some circumstances also cause DNA cleavage. End-filling is a gentler method that rarely causes breakage of the DNA, but unfortunately can only be used to label DNA molecules that have sticky ends. The enzyme used is the Klenow fragment (p. 55), which will 'fill in' a sticky end by synthesizing the complementary strand (Figure 4.15c). As with nick translation, if the end-filling reaction is carried out in the presence of labelled nucleotides, then the DNA itself will become labelled.

Both nick translation and end-filling can label DNA to such an extent that very small quantities can be detected in gels by autoradiography. As little as 2 ng of DNA per band can be visualized under ideal conditions.

(a) [α – ³²P] dATP

(b) Labelling by nick translation

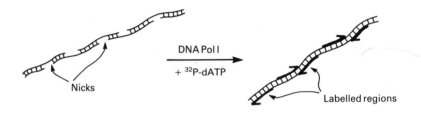

(c) Labelling by end-filling

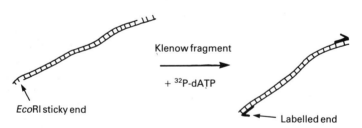

Figure 4.15 Radioactive labelling. (a) The structure of [α – ³²P]dATP. (b) Labelling DNA by nick translation. (c) Labelling DNA by end-filling.

4.2.7 ESTIMATION OF THE SIZES OF DNA MOLECULES

Gel electrophoresis will separate different-sized DNA molecules, with the smallest molecules travelling the greatest distance towards the positive electrode. If several DNA fragments of varying sizes are present (the result of a successful restriction digest, for example), then a series of bands will appear in the gel. How can the sizes of these fragments be determined?

The most accurate method is to make use of the mathematical relationship that links migration rate to molecular weight. The relevant formula is

$$D = a - b(\log M)$$

where D is the distance moved, M is the molecular weight, and a and b are constants that depend on the electrophoresis conditions.

However, a much simpler though less precise way of estimating DNA fragment sizes is generally used. A standard restriction digest, comprising fragments of known size, is usually included in each electrophoresis gel that is run. Restriction digests of λ DNA are often used in this way as size markers. For example, HindIII cleaves λ DNA into eight fragments, ranging in size from 125 bp for the smallest to over 23 kb for the largest. As the sizes of the fragments in this digest are known, the fragment sizes in the experimental digest can be estimated by comparing the positions of the bands in the two tracks (Figure 4.16). Although not precise, this method can be performed with as little as a 5% error, which is quite satisfactory for most purposes.

4.2.8 MAPPING THE POSITIONS OF DIFFERENT RESTRICTION SITES ON A DNA MOLECULE

So far we have considered how to determine the number and sizes of the DNA fragments produced by restriction endonuclease cleavage. The next step in **restriction analysis** is to construct a map showing the relative positions on the DNA molecule of the recognition sequences for a number of different enzymes. Only when a **restriction map** is available can the correct restriction endonucleases be selected for the particular cutting manipulation that is required (Figure 4.17).

To construct a restriction map, a series of restriction digests must be performed. Firstly, the number and sizes of the fragments produced by each restriction endonuclease must be determined by gel electrophoresis, followed by comparison with size markers (Figure 4.18). This information must then be supplemented by a series of **double digestions**, in which the DNA is cut by two restriction endonucleases at once. It may be possible to perform a double digestion in one step, if both enzymes have similar requirements for pH, Mg^{2+} concentration, etc. Alternatively, the two digestions may have to be carried out one after the other, adjusting the reaction mixture after the first digestion to provide a different set of conditions for the second enzyme.

Comparing the results of single and double digests will allow many, if not all, of the restriction sites to be mapped (Figure 4.18). Ambiguities can usually be resolved by **partial digestion**, carried out under conditions that will result in cleavage of only a limited number of the restriction sites on any DNA molecule. Partial digestion is achieved normally either by a short incubation period, so the enzyme does not have time to cut all the restriction

(a) Rough estimation by eye

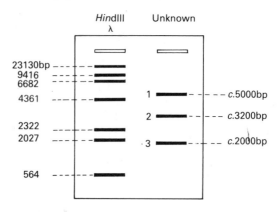

(b) Accurate graphical estimation

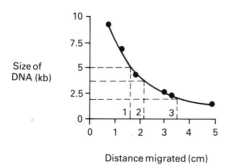

Distance migrated (cm)

Figure 4.16 Estimation of the sizes of DNA fragments in an agarose gel. (a) A rough estimate of fragment size can be obtained by eye. (b) A more accurate measurement of fragment size is gained by using the mobilities of the *Hind*III-fragments to construct a calibration curve; the sizes of the unknown fragments can then be determined from the distances they have migrated.

sites, or by incubation at a low temperature (e.g. 4°C rather than 37°C), which will limit the activity of the enzyme. The result of a partial digestion will be a complex pattern of bands in an electrophoresis gel. As well as the standard fragments, produced by total digestion, additional sizes will be seen. These are molecules that comprise two adjacent restriction fragments, separated by a site that has not been cleaved. Their sizes will indicate which restriction fragments in the complete digest are next to one another in the uncut molecule (Figure 4.18).

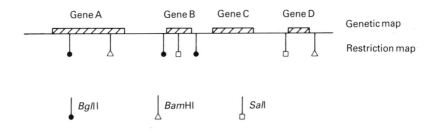

To obtain gene B, digest with *Bgl* II

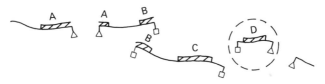

To obtain gene D, digest with *Bam*HI + *Sal*I

Figure 4.17 Using a restriction map to work out which restriction endonucleases should be used to obtain DNA fragments containing individual genes.

4.3 LIGATION – JOINING DNA MOLECULES TOGETHER

The final step in construction of a recombinant DNA molecule is the joining together of the vector molecule and the DNA to be cloned (Figure 4.19). This process is referred to as ligation, and the enzyme that catalyses the reaction is called DNA ligase.

4.3.1 THE MODE OF ACTION OF DNA LIGASE

All living cells produce DNA ligases, but the enzyme used in genetic engineering is usually purified from *E. coli* bacteria that have been infected with T4 phage. Within the cell the enzyme carries out the very important function of repairing any discontinuities that may arise in one of the strands of a double-stranded molecule (Figure 4.4a). A discontinuity is quite simply a position where a phosphodiester bond between adjacent nucleotides is

Single and double digestions

Enzyme	Number of fragments	Sizes (kb)
XbaI	2	24.0. 24.5
XhoI	2	15.0. 33.5
KpnI	3	1.5. 17.0. 30.0
XbaI + XhoI	3	9.0. 15.0. 24.5
XbaI + KpnI	4	1.5. 6.0. 17.0. 24.0

Conclusions: 1. As λ DNA is linear, the number of restriction sites for each enzyme is XbaI 1, XhoI 1, KpnI 2.

2. The XbaI and XhoI sites can be mapped:

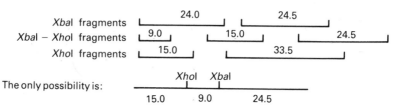

The only possibility is:

3. All the KpnI sites fall in the 24.5 kb XbaI fragment, as the 24.0 kb fragment is intact after XbaI — KpnI double digestion. The order of the KpnI fragments can only be determined by partial digestion.

Partial digestion

Enzyme	Fragment sizes (kb)
KpnI — limiting conditions	1.5, 17.0, 18.5, 30.0, 31.5, 48.5

Conclusions: 48.5 kb fragment = uncut λ.
1.5. 17.0 and 30.0 kb fragments are products of complete digestion.
18.5 and 31.5 kb fragments are products of partial digestion.

The KpnI map must be:

Therefore the complete map is:

Figure 4.18 Restriction mapping. This example shows how the positions of the XbaI, XhoI and KpnI sites on the λ DNA molecule can be determined.

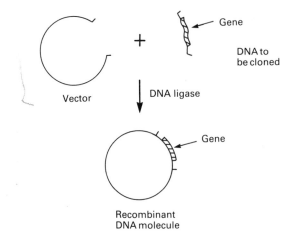

Figure 4.19 Ligation: the final step in construction of a recombinant DNA molecule.

missing (contrast this with a 'nick', where one or more nucleotides are absent). Although discontinuities may arise by chance breakage of the cell's DNA molecules, they are also a natural result of processes such as DNA replication and recombination. Ligases therefore play several vital roles in the cell.

In the test-tube, purified DNA ligases, as well as repairing single-strand discontinuities, will also join together individual DNA molecules or the two ends of the same molecule. The chemical reaction involved in ligating two molecules is exactly the same as discontinuity repair, except that two phosphodiester bonds must be made, one for each strand (Figure 4.20a).

4.3.2 STICKY ENDS INCREASE THE EFFICIENCY OF LIGATION

The ligation reaction in Figure 4.20a shows two blunt-ended fragments being joined together. Although this reaction can be carried out in the test-tube, it is not very efficient. This is because the ligase is unable to 'catch hold' of the molecule to be ligated, and has to wait for chance associations to bring the ends together. If possible, blunt-end ligation should be performed at high DNA concentrations, to increase the chances of the ends of the molecules coming together in the correct way.

In contrast, ligation of complementary sticky ends is much more efficient. This is because compatible sticky ends can base pair with one another by hydrogen bonding (Figure 4.20b) forming a relatively stable structure for the enzyme to work on. If the phosphodiester bonds are not synthesized fairly quickly, then the sticky ends will fall apart again. These transient, base-paired structures do, however, increase the efficiency of ligation by increasing the length of time the ends are in contact with one another.

(a) Ligating blunt ends

(b) Ligating sticky ends

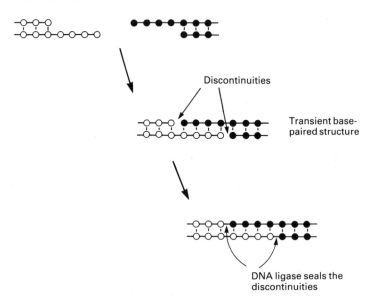

Discontinuities

Transient base-paired structure

DNA ligase seals the discontinuities

Figure 4.20 The different joining reactions catalysed by DNA ligase. (a) Ligation of blunt-ended molecules. (b) Ligation of sticky-ended molecules.

4.3.3 PUTTING STICKY ENDS ON TO A BLUNT-ENDED MOLECULE

For the reasons detailed in the preceding section, compatible sticky ends are desirable on the DNA molecules to be ligated together in a gene cloning experiment. Often these sticky ends can be provided by digesting both the vector and the DNA to be cloned with the same restriction endonuclease, or with different enzymes that produce the same sticky end. However, it is not always possible to do this. A common situation is where the vector molecule has sticky ends, but the DNA fragments to be cloned are blunt-ended. Under these circumstances one of three methods can be used to put the correct sticky ends on to the DNA fragments.

like those from PCR

(a) **A typical linker**

C-G-A-T-G-G-A-T-C-C-A-T-C-G
| | | | | | | | | | | | | |
G-C-T-A-C-C-T-A-G-G-T-A-G-C

*Bam*HI site

(b) **The use of linkers**

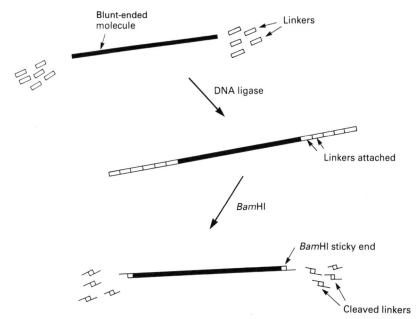

Figure 4.21 Linkers and their use. (a) The structure of a typical linker. (b) The attachment of linkers to a blunt-ended molecule.

(a) Linkers

The first of these methods involves the use of **linkers**. These are short pieces of double-stranded DNA, of known nucleotide sequence, that are synthesized in the test-tube. A typical linker is shown in Figure 4.21a. It is blunt-ended, but contains a restriction site, *Bam*HI in the example shown. DNA ligase will attach linkers to the ends of larger blunt-ended DNA molecules. Although a blunt-end ligation, this particular reaction can be performed very efficiently because synthetic oligonucleotides, such as linkers, can be made in very large amounts and added into the ligation mixture at a high concentration.

More than one linker will attach to each end of the DNA molecule, producing the chain structure shown in Figure 4.21b. However, digestion with *Bam*HI will cleave the chains at the recognition sequences, producing a large number of cleaved linkers and the original DNA fragment, now carrying *Bam*HI sticky ends. This modified fragment is ready for ligation into a cloning vector restricted with *Bam*HI.

(b) Adaptors

The second method of attaching a sticky end to a blunt-ended molecule is through the use of **adaptors**. Adaptors, like linkers, are short synthetic oligonucleotides. However, unlike linkers, an adaptor is synthesized so that it has one blunt end and one sticky end. The idea is of course to ligate the blunt end of the adaptor to the blunt ends of the DNA fragment, to produce a new molecule with sticky ends (Figure 4.22a).

This may appear to be a simpler method than attachment of linkers, but there is a complication with the use of adaptors. The sticky ends of individual adaptor molecules will of course base-pair with each other to form dimers (Figure 4.22b), so that the new DNA molecule will still be blunt-ended (Figure 4.22c). The sticky end could be recreated by digestion with a restriction endonuclease, as with linkers, but this would defeat the purpose of synthesizing adaptors with sticky ends.

The answer to the problem lies in the precise chemical structure of the ends of the adaptor molecule. Normally, the two ends of a polynucleotide strand are chemically distinct, a fact that will be clear from a careful examination of the polymeric structure (Figure 4.23a). One end, referred to as the 5'-terminus, carries a phosphate group (5'-P); the other, the 3'-terminus, has a hydroxyl group (3'-OH). In the double helix the two strands are antiparallel (Figure 4.23b), so each end of a double-stranded molecule consists of one 5'-P terminus and one 3'-OH terminus. Ligation normally takes place between the 5'-P and 3'-OH ends (Figure 4.23c).

Adaptor molecules are synthesized so that the blunt end is the same as 'natural' DNA, but the sticky end is different. The 3'-OH terminus of the sticky end is the same as usual, but the 5'-P terminus is modified; it lacks the phosphate group, and is in fact a 5'-OH terminus (Figure 4.24a). DNA ligase is unable to form a phosphodiester bridge between 5'-OH and 3'OH ends. The result is that, although base pairing is always occurring between the sticky ends of adaptor molecules, the association is never stabilized by ligation (Figure 4.24b).

Adaptors can therefore be ligated to a DNA molecule but not to themselves. After the adaptors have been attached the abnormal 5'-OH terminus is converted to the natural 5'-P form by treatment with the enzyme polynucleotide kinase (p. 56), producing a sticky-ended fragment that can be inserted into an appropriate vector.

(c) Producing sticky ends by homopolymer tailing

The technique of **homopolymer tailing** offers a radically different approach to the production of sticky ends on a blunt-ended DNA molecule. A

(a) A typical adaptor

G-A-T-C-C-C-G-G
       | | | |
       G-G-C-C

*Bam*HI sticky
end

(b) Adaptors could ligate to one another

C - C - G - G G - A - T - C - C - C - G - G
G - G - C - C - C - T - A - G G - G - C - C

(c) The new DNA molecule is still blunt ended

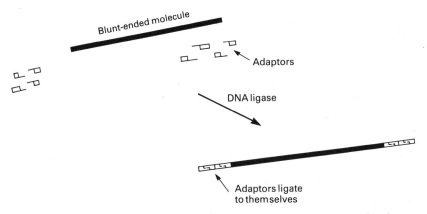

Blunt-ended molecule

Adaptors

DNA ligase

Adaptors ligate
to themselves

Figure 4.22 Adaptors and the problems with their use. (a) A typical adaptor. (b) Adaptors can ligate to one another to produce a molecule similar to a linker, so that (c) after ligation of adaptors a blunt-ended molecule is still blunt-ended.

homopolymer is simply a polymer in which all the subunits are the same. A DNA strand made up entirely of, say, deoxyguanosine is an example of a homopolymer, and is referred to as polydeoxyguanosine or poly(dG).

Tailing involves using the enzyme terminal deoxynucleotidyl transferase (p. 56) to add a series of nucleotides on to the 3'-OH termini of a double-stranded DNA molecule. If this reaction is carried out in the presence of just one deoxynucleotide, then a homopolymer tail will be produced (Figure 4.25a).

Of course, to be able to ligate together two tailed molecules, the

(a) The structure of the polynucleotide strand showing the chemical distinction between the 5′-P and 3′-OH termini

One nucleotide

Base

Base

Base

OH
3′

5′

(b) In the double helix the polynucleotide strands are antiparallel

5′ →→→→→→→→→ 3′
3′ ←←←←←←←←← 5′

(c) Ligation takes place between 5′-P and 3′-OH termini

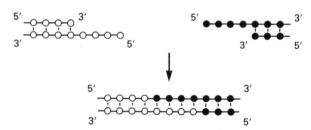

Figure 4.23 The distinction between the 5′-and 3′-termini of a polynucleotide.

(a) The precise structure of an adaptor

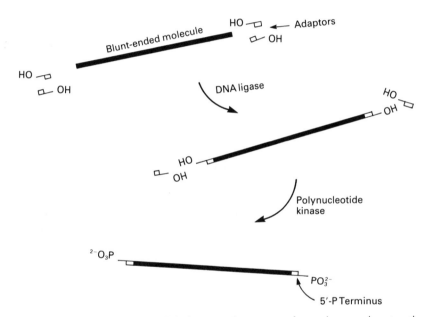

(b) Ligation using adaptors

Figure 4.24 The use of adaptors. (a) The actual structure of an adaptor, showing the modified 5'-OH terminus. (b) Conversion of blunt ends to sticky ends through the attachment of adaptors.

homopolymers must be complementary. Frequently poly(dG) tails are attached to the vector and poly(dC) to the DNA to be cloned. Base pairing between the two will occur when the DNA molecules are mixed (Figure 4.25b).

In practice the poly(dG) and poly(dC) tails are not usually exactly the same length, and base-paired recombinant molecules will have nicks as well as discontinuities (Figure 4.25c). Repair is therefore a two-step process, using Klenow polymerase to fill in the nicks followed by DNA ligase to synthesize the final phosphodiester bonds. This repair reaction does not always have to

(a) Synthesis of a homopolymer tail

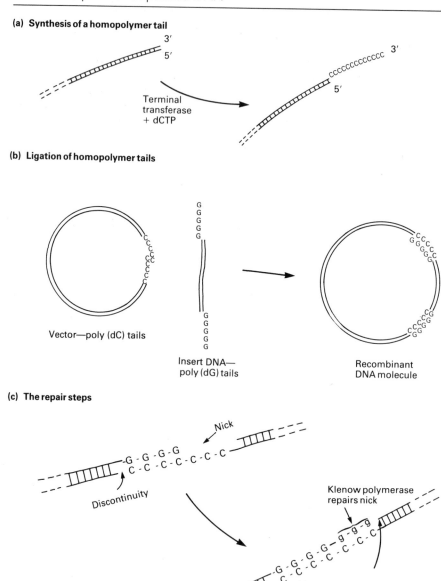

(b) Ligation of homopolymer tails

Vector—poly (dC) tails

Insert DNA—
poly (dG) tails

Recombinant
DNA molecule

(c) The repair steps

Figure 4.25 Homopolymer tailing. (a) Synthesis of a homopolymer tail. (b) Construction of a recombinant DNA molecule from a tailed vector plus tailed insert DNA. (c) Repair of the recombinant DNA molecule.

be performed in the test-tube. If the complementary homopolymer tails are longer than about 20 nucleotides, then quite stable base-paired associations will be formed. A recombinant DNA molecule, held together by base pairing although not completely ligated, will often be stable enough to be introduced into the host cell in the next stage of the cloning experiment (Figure 1.1). Once inside the host, the cell's own DNA polymerase and DNA ligase will repair the recombinant DNA molecule, completing the construction begun in the test-tube.

FURTHER READING

Smith, H. O. and Wilcox, K. W. (1970) A restriction enzyme from *Haemophilus influenzae. Journal of Molecular Biology*, **51**, 379–91 – one of the first full descriptions of a restriction endonuclease.

Roberts, R. J. (1989) Restriction enzymes and their isoschizomers. *Nucleic Acids Research*, **17**, r347–87 – a comprehensive list of all the known restriction endonucleases and their recognition sequences.

Lobban, P. and Kaiser, A. D. (1973) Enzymatic end-to-end joining of DNA molecules. *Journal of Molecular Biology*, **79**, 453–71 – ligation.

Rothstein, R. J. *et al.* (1979) Synthetic adaptors for cloning DNA. *Methods in Enzymology*, **68**, 98–109.

Introduction of DNA into living cells

5

The manipulations described in Chapter 4 allow the molecular biologist to create novel recombinant DNA molecules. The next step in a gene cloning experiment is to introduce these molecules into living cells, usually bacteria, which will then grow and divide to produce clones (Figure 1.1). Strictly speaking, the word 'cloning' refers only to the later stages of the procedure, and not to the construction of the recombinant DNA molecule itself.

Cloning serves two main purposes. Firstly, it allows a large number of recombinant DNA molecules to be produced from a limited amount of starting material. At the outset only a few nanograms of recombinant DNA may be available, but each bacterium that takes up a plasmid will divide numerous times to produce a colony, each cell of which will contain multiple copies of the molecule. Several micrograms of recombinant DNA can usually be prepared from a single bacterial colony, representing a thousandfold increase over the starting amount (Figure 5.1). If the colony is used not as a source of DNA, but as an inoculum for a liquid culture, then the resulting cells may provide milligrams of DNA, a millionfold increase in yield. In this way cloning can supply the large amounts of DNA needed for molecular biological studies of gene structure and expression (Chapters 9 and 10).

The second important function of cloning can be described as purification. The manipulations that result in a recombinant DNA molecule can only rarely be controlled to the extent that no other DNA molecules are present at the end of the procedure. The ligation mixture may contain, in addition to the desired recombinant molecule, any number of the following (Figure 5.2a):

1. unligated vector molecules;
2. unligated DNA fragments;
3. vector molecules that have recircularized without new DNA being inserted ('self-ligated' vector);
4. recombinant DNA molecules that carry the wrong inserted DNA fragment.

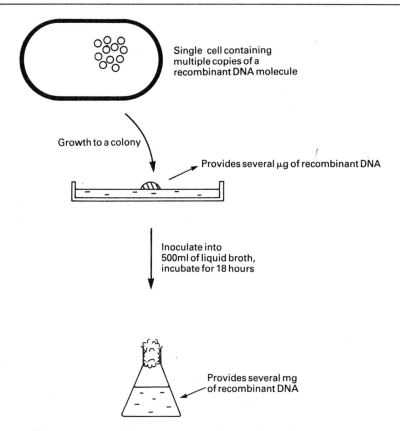

Single cell containing
multiple copies of a
recombinant DNA molecule

Growth to a colony

Provides several µg of recombinant DNA

Inoculate into
500ml of liquid broth,
incubate for 18 hours

Provides several mg
of recombinant DNA

Figure 5.1 Cloning can supply large amounts of recombinant DNA.

Unligated molecules will rarely cause a problem because, even though they may be taken up by bacterial cells, only under exceptional circumstances will they be replicated. It is much more likely that the host enzymes will degrade these pieces of DNA. On the other hand, self-ligated vector molecules and incorrect recombinant plasmids will be replicated just as efficiently as the desired molecule (Figure 5.2b). However, purification of the desired molecule can still be achieved through cloning because it will be extremely unusual for any one cell to take up more than one DNA molecule. Each cell gives rise to a single colony, so each of the resulting clones will consist of cells that all contain the same molecule. Of course different colonies will contain different molecules: some will contain the desired recombinant DNA molecule, some will have different recombinant molecules, and some self-ligated vector. The problem therefore becomes a question of identifying the colonies that contain the correct recombinant plasmids.

This chapter is concerned with the way in which plasmid and phage

(a) The products of ligation

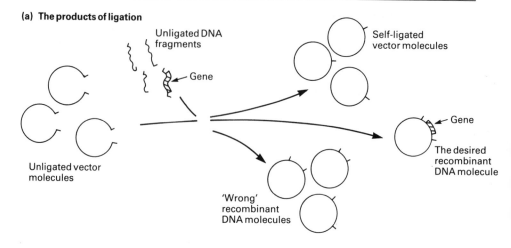

(b) All circular molecules will be cloned

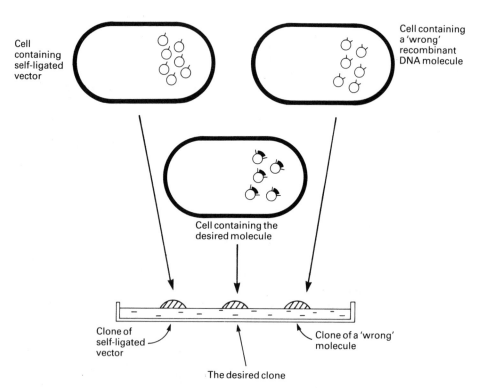

Figure 5.2 Cloning is analogous to purification. From a mixture of different molecules, clones containing copies of just one molecule can be obtained.

vectors, and recombinant molecules derived from them, are introduced into bacterial cells. During the course of the chapter it will become apparent that selection for colonies containing recombinant molecules, as opposed to colonies containing self-ligated vector, is relatively easy. The more difficult proposition of how to distinguish clones containing the correct recombinant DNA molecule from all the other recombinant clones will be tackled in Chapter 8.

5.1 TRANSFORMATION – THE UPTAKE OF DNA BY BACTERIAL CELLS

Most species of bacteria are able to take up DNA molecules from the medium in which they grow. Often a DNA molecule taken up in this way will be degraded, but occasionally it will be able to survive and replicate in the host cell. In particular this will happen if the DNA molecule is a plasmid with an origin of replication recognized by the host.

Uptake and stable retention of a plasmid is usually detected by looking for expression of the genes carried by the plasmid (p. 89). For example, *E. coli* cells are normally sensitive to the growth inhibitory effects of the antibiotics ampicillin and tetracycline. However, cells that contain the plasmid pBR322 (p. 104), a popular cloning vector, are resistant to these antibiotics. This is because pBR322 carries two sets of genes, one gene that codes for a β-lactamase enzyme that modifies ampicillin into a form that is non-toxic to the bacterium, and a different set of genes that code for enzymes that detoxify tetracycline. Uptake of pBR322 can be detected because the *E. coli* cells are **transformed** from ampicillin and tetracycline sensitive (amp^{S}tet^S) to ampicillin and tetracycline resistant (amp^{R}tet^R).

In recent years the term **transformation** has been extended to include uptake of any DNA molecule by any type of cell, regardless of whether the uptake results in a detectable change in the cell, or whether the cell involved is bacterial, fungal, animal or plant.

5.1.1 NOT ALL SPECIES OF BACTERIA ARE EQUALLY EFFICIENT AT DNA UPTAKE

In nature, transformation is probably not a major process by which bacteria obtain genetic information. This is reflected by the fact that in the laboratory only a few species (notably members of the genera *Bacillus* and *Streptococcus*) can be transformed with ease. Close study of these organisms has revealed that they possess sophisticated mechanisms for DNA binding and uptake.

Most species of bacteria, including *E. coli*, take up only limited amounts of DNA under normal circumstances. In order to transform these species efficiently, the bacteria have to undergo some form of physical and/or chemical treatment that will enhance their ability to take up DNA. Cells that have undergone this treatment are said to be **competent**.

5.1.2 PREPARATION OF COMPETENT *E. coli* CELLS

As with many breakthroughs in recombinant DNA technology, the key development as far as transformation was concerned occurred in the early 1970s, when it was observed that *E. coli* cells that had been soaked in an ice-cold salt solution were more efficient at DNA uptake than unsoaked cells. A solution of 50 mM calcium chloride is traditionally used, although other salts, notably rubidium chloride, are also effective.

Exactly why this treatment works is not understood. Possibly $CaCl_2$ causes the DNA to precipitate on to the outside of the cells, or perhaps the salt is responsible for some kind of change in the cell wall that improves DNA binding. In any case, soaking in $CaCl_2$ affects only DNA binding, and not the actual uptake into the cell. When DNA is added to treated cells, it remains attached to the cell exterior, and is not at this stage transported into the cytoplasm (Figure 5.3). The actual movement of DNA into competent cells is stimulated by briefly raising the temperature to 42°C. Once again, the exact reason why this heat-shock is effective is not understood.

5.1.3 SELECTION FOR TRANSFORMED CELLS

Transformation of competent cells is an inefficient procedure, however carefully the cells have been prepared. Although 1 ng of pBR322 can yield

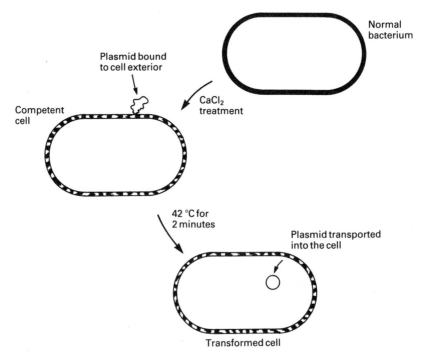

Figure 5.3 The binding and uptake of DNA by a competent bacterial cell.

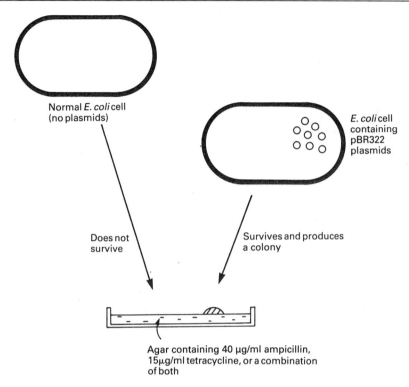

Figure 5.4 Selecting cells that contain pBR322 plasmids by plating on to agar medium containing ampicillin and/or tetracycline.

1000–10 000 transformants, this represents the uptake of only 0.01% of all the available molecules. Furthermore, 10 000 transformants is only a very small proportion of the total number of cells that will be present in a competent culture. This last fact means that some way must be found to distinguish a cell that has taken up a plasmid from the many thousands that have not been transformed.

The answer is to make use of a selectable marker carried by the plasmid. In the case of pBR322, reference to Figure 5.4 will show that only *E. coli* cells that carry the plasmid will be able to form colonies on an agar medium that contains ampicillin or tetracycline. Most plasmid cloning vectors carry at least one gene that confers antibiotic resistance on the host cells, and selection for transformants is achieved simply by plating on to an agar medium that contains the relevant antibiotic.

Resistance depends on the presence in the transformed cells of the enzyme, coded by the plasmid, that detoxifies the antibiotic. If the transformed cells are plated on to the selective medium immediately after the heat-shock treatment, then only a limited number of the enzyme molecules will be present – the cells have only recently taken up a plasmid

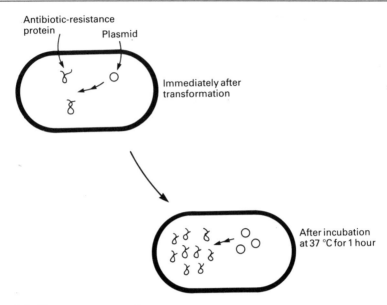

Figure 5.5 Phenotypic expression.

and expression of plasmid genes will only just be underway. Before plating, the cells are therefore placed in a small volume of liquid medium, in the absence of antibiotic, and incubated for a short time. This allows plasmid replication and expression to get started, so that when the cells are plated out and encounter the antibiotic, they will already have synthesized sufficient resistance enzymes to be able to survive (Figure 5.5).

5.2 SELECTION FOR RECOMBINANTS

A **recombinant** is a transformed cell that harbours a plasmid that is a recombinant DNA molecule. In a gene cloning experiment it is almost always necessary to be able to select these recombinant cells from transformed cells that contain normal plasmids.

5.2.1 INSERTIONAL INACTIVATION OF A SELECTABLE MARKER

Most cloning vectors are designed so that insertion of a DNA fragment into the plasmid destroys the integrity of one of the genes present on the molecule. The general principles of **insertional inactivation** are illustrated by a typical cloning experiment using pBR322 as the vector.

pBR322 has several unique restriction sites that can be used to open up the vector prior to insertion of a new DNA fragment (Figure 5.6a). *Bam*HI, for

(a) The normal vector molecule

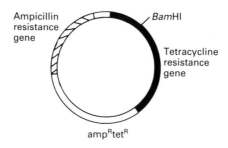

Ampicillin
resistance
gene

BamHI

Tetracycline
resistance
gene

amp^Rtet^R

(b) A recombinant pBR322 molecule

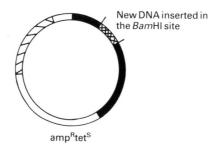

New DNA inserted in
the BamHI site

amp^Rtet^S

Figure 5.6 The cloning vector pBR322. (a) The normal vector molecule. (b) A recombinant molecule containing an extra piece of DNA inserted into the BamHI site. For a more detailed map of pBR322 see Figure 6.1

example, cuts pBR322 at just one point, within the cluster of genes that code for resistance to tetracycline. A recombinant pBR322 molecule, one that carries an extra piece of DNA in the BamHI site (Figure 5.6b), will therefore no longer be able to confer tetracycline resistance on its host, as one of the necessary genes is now disrupted by the inserted DNA. Cells containing this recombinant pBR322 will still be resistant to ampicillin, but sensitive to tetracycline (amp^{R}tet^S).

Screening for pBR322 recombinants is performed in the following way. After transformation the cells are plated on to ampicillin medium and incubated until colonies appear (Figure 5.7a). All colonies will consist of transformants (remember, untransformed cells are amp^S so do not produce colonies), but only a few will contain recombinant pBR322 molecules, most will contain the normal plasmid. To distinguish recombinants the colonies are **replica-plated** on to agar medium that contains tetracycline (Figure 5.7b). After incubation, some of the original colonies will regrow, others will

(a) Colonies on ampicillin medium

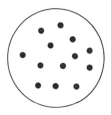

(b) Replica plating

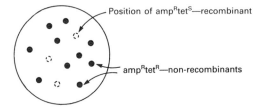

(c) Amp^R tet^R colonies grown on tetracycline medium

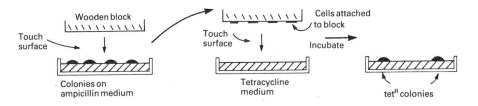

Figure 5.7 Screening for pBR322 recombinants by insertional inactivation of the tetracycline resistance gene. (a) Cells are plated on to ampicillin agar: all the transformants produce colonies. (b) The colonies are replica-plated on to tetracycline medium. (c) The colonies that grow on tetracycline medium are amp^R tet^R and therefore non-recombinants. Recombinants (amp^R tet^S) do not grow, but their position on the ampicillin plate is now known.

not (Figure 5.7c). Those that do will consist of cells that carry the normal pBR322 with no inserted DNA and therefore a functional tetracyline resistance gene cluster (amp^R tet^R). On the other hand, colonies on the ampicillin plate that do not grow on the tetracycline plate will consist of recombinant cells (amp^R tet^S).

5.2.2 INSERTIONAL INACTIVATION DOES NOT ALWAYS INVOLVE ANTIBIOTIC RESISTANCE

Although an antibiotic resistance gene is the most convenient type of selectable marker, some plasmids make use of different systems. An important example is pUC8 (Figure 5.8a), which carries the ampicillin resistance gene and a gene called *lacZ'*, which codes for a part of the enzyme β-galactosidase. Cloning with pUC8 involves insertional inactivation of the *lacZ'* gene; recombinants are distinguished by their inability to synthesize β-galactosidase (Figure 5.8b).

β-galactosidase is one of a series of enzymes involved in the breakdown of lactose to. glucose plus galactose. It is normally coded by the gene *lacZ*, which resides on the *E. coli* chromosome. Some strains of *E. coli* have a modified *lacZ* gene, one that lacks a large segment referred to as *lacZ'* and coding for the α-peptide portion of β-galactosidase (Figure 5.9a). These

(a) pUC8

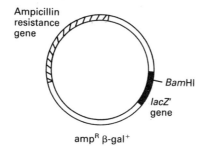

amp^R β-gal$^+$

(b) A recombinant pUC8 molecule

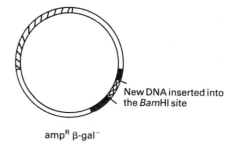

amp^R β-gal$^-$

Figure 5.8 The cloning vector pUC8. (a) The normal vector molecule. (b) A recombinant molecule containing an extra piece of DNA inserted into the *Bam*HI site. For a more detailed map of pUC8 see Figures 6.3 and 6.4.

(a) **The role of the *lacZ'* gene**

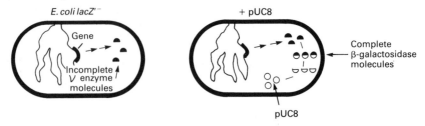

- β-galactosidase fragment coded by bacterial gene
- β-galactosidase fragment coded by plasmid gene
- Complete β-galactosidase molecule

(b) **Screening for pUC8 recombinants**

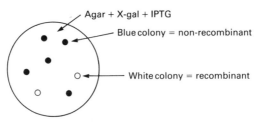

Blue colonies = β-galactosidase synthesized
X-gal → blue product

White colonies = β-galactosidase not synthesized
X-gal → no blue product

Figure 5.9 The rationale behind insertional inactivation of the *lacZ'* gene carried by pUC8. (a) The bacterial and plasmid genes complement each other to produce a functional β-galactosidase molecule. (b) Recombinants are screened by plating on to agar containing X-gal plus IPTG.

mutants can synthesize the enzyme only when they harbour a plasmid, such as pUC8, that carries the missing *lacZ'* segment of the gene.

A cloning experiment with pUC8 involves selection of transformants on ampicillin agar followed by screening for β-galactosidase activity to distinguish recombinants. Cells that harbour a normal pUC8 plasmid will be amp^R and able to synthesize β-galactosidase (Figure 5.8a); recombinants will also be amp^R but unable to make β-galactosidase (Figure 5.8b).

Screening for β-galactosidase presence or absence is in fact quite straightforward. Rather than assay for lactose being split to glucose and galactose, a slightly different reaction catalysed by the enzyme is tested for.

This involves a lactose analogue called X-gal (5-bromo-4-chloro-3-indolyl-β-D-galactopyranoside) which is broken down by β-galactosidase to a product that is coloured deep blue. If X-gal (plus an inducer of the enzyme such as isopropyl-thiogalactoside, IPTG) is added to the agar, along with ampicillin, then non-recombinant colonies, the cells of which synthesize β-galactosidase, will be coloured blue, whereas recombinants with a disrupted *lacZ'* gene and unable to make β-galactosidase, will be white. This scheme is summarized in Figure 5.9b.

5.3 INTRODUCTION OF PHAGE DNA INTO BACTERIAL CELLS

There are two different methods by which a recombinant DNA molecule constructed with a phage vector can be introduced into a bacterial cell: transfection and *in vitro* packaging.

(a) Transfection

This process is equivalent to transformation, the only difference being that phage DNA rather than a plasmid is involved. Just as with a plasmid, the purified phage DNA, or recombinant phage molecule, is mixed with competent *E. coli* cells and DNA uptake induced by heat-shock.

(b) *In vitro* packaging

Transfection with λ DNA molecules is not a very efficient process when compared with the infection of a culture of cells with mature λ phage particles. It would therefore be useful if recombinant λ molecules could be packaged into the λ head-and-tail structures in the test-tube.

This may sound difficult but in fact is relatively easy to achieve. Packaging requires a number of different proteins coded by the λ genome, but these can be prepared at a high concentration from cells infected with defective λ phage strains. Two different systems are in use. With the single-strain system the defective λ phage carries a mutation in the *cos* sites, so that these are not recognized by the endonuclease that normally cleaves the λ concatamers during phage replication (p. 22). The defective phage cannot therefore replicate, though it does direct synthesis of all the proteins needed for packaging. The proteins accumulate in the bacterium and can be purified from cultures of *E. coli* infected with the mutated λ, and used for *in vitro* packaging of recombinant λ molecules (Figure 5.10a).

With the second system two defective λ strains are needed. These strains carry a mutation in one of the components of the phage protein coat. Infected cells will continue to synthesize and accumulate all the other components, even though mature phage particles cannot be assembled (Figure 5.10b). An *in vitro* packaging mix can therefore be prepared by

(a) A single-strain packaging system

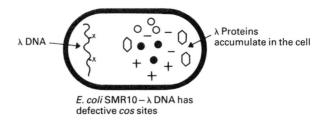

λ DNA →

λ Proteins accumulate in the cell

E. coli SMR10 – λ DNA has defective *cos* sites

(b) A two-strain packaging system

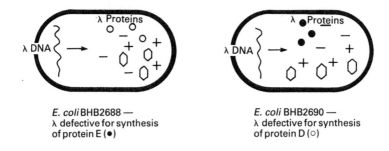

λ Proteins

λ DNA →

E. coli BHB2688 — λ defective for synthesis of protein E (●)

λ Proteins

λ DNA →

E. coli BHB2690 — λ defective for synthesis of protein D (○)

(c) *In vitro* packaging

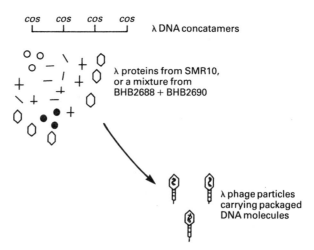

cos   *cos*   *cos*   *cos*

λ DNA concatamers

λ proteins from SMR10, or a mixture from BHB2688 + BHB2690

λ phage particles carrying packaged DNA molecules

Figure 5.10 *In vitro* packaging. (a) Synthesis of λ capsid proteins by *E. coli* strain SMR10, which carries a λ phage that has defective *cos* sites. (b) Synthesis of incomplete sets of λ capsid proteins by *E. coli* strains BHB2688 and BHB2690. (c) A mixture of cell lysates provides the complete set of capsid proteins and can package λ DNA molecules in the test-tube.

mixing lysates of two cultures of cells, one infected with a λ strain defective in one protein, and one defective in a second protein. The mixture will contain all the necessary components and will package recombinant λ molecules into mature phage particles (Figure 5.10c). With both systems, packaged molecules can be introduced into *E. coli* cells simply by adding the assembled phage to the bacterial culture and allowing the normal λ infective process to take place.

5.3.1 PHAGE INFECTION IS VISUALIZED AS PLAQUES ON AN AGAR MEDIUM

The final-stage of the phage infection cycle is cell lysis (p. 18). If infected cells are spread on to a solid agar medium immediately after addition of the phage, or immediately after transfection with phage DNA, then cell lysis can be visualized as **plaques** on a lawn of bacteria (Figure 5.11a). Each plaque is a zone of clearing produced as the phage lyse the cells and move on to infect and eventually lyse the neighbouring bacteria (Figure 5.11b).

Both λ and M13 form plaques. With λ these are true plaques, produced by cell lysis. However, M13 plaques are slightly different as M13 does not lyse the host cells (p. 19). Instead M13 causes a decrease in the growth rate of infected cells sufficient to produce a zone of relative clearing on a bacterial lawn. Although not a true plaque, these zones of clearing are visually identical to normal phage plaques (Figure 5.11c).

The end result of a gene cloning experiment using a λ or M13 vector is therefore an agar plate covered in phage plaques. Each plaque is derived from a single transfected or infected cell and therefore contains identical phage particles. These may contain self-ligated vector molecules, or they may be recombinants.

5.4 SELECTION FOR RECOMBINANT PHAGE

As most phage lyse the cells that they infect, any form of selection based on antibiotic resistance would be totally inappropriate. Insertional inactivation of an antibiotic resistance gene, although the most popular form of recombinant selection for plasmid vectors, cannot therefore be used for phage vehicles. Instead, a variety of ways of distinguishing recombinant plaques have been developed. The following are the most important.

5.4.1 INSERTIONAL INACTIVATION OF A *lacZ'* GENE CARRIED BY THE PHAGE VECTOR

All M13 cloning vectors (p. 110), as well as a few λ vectors, carry a copy of the *lacZ'* gene. Insertion of new DNA into this gene inactivates β-galactosidase synthesis, just as with the plasmid vector pUC8. Recombinants are distinguished by plating cells on to X-gal agar: plaques comprising normal phage are blue, recombinant plaques are clear (Figure 5.12a).

(a) Plaques on a lawn of bacteria

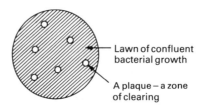

Lawn of confluent
bacterial growth

A plaque – a zone
of clearing

(b) Lytic plaques

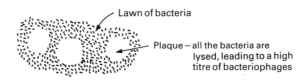

Lawn of bacteria

Plaque – all the bacteria are
lysed, leading to a high
titre of bacteriophages

(c) M13 plaques

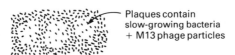

Plaques contain
slow-growing bacteria
+ M13 phage particles

Figure 5.11 Bacteriophage plaques. (a) The appearance of plaques on a lawn of bacteria. (b) Plaques produced by a phage that lyses the host cell (e.g. λ in the lytic infection cycle); the plaques contain lysed cells plus many phage particles. (c) Plaques produced by M13; these plaques contain slow-growing bacteria plus many M13 phage particles.

5.4.2 INSERTIONAL INACTIVATION OF THE λ *cI* GENE

Several types of λ cloning vector have unique restriction sites in the *cI* gene (map position 38 on Figure 2.9). Insertional inactivation of this gene causes a change in plaque morphology. Normal plaques appear 'turbid', whereas recombinants with a disrupted *cI* gene are 'clear' (Figure 5.12b). The difference is readily apparent to the experienced eye.

5.4.3 SELECTION USING THE SPI PHENOTYPE

λ phage cannot normally infect *E. coli* cells that already possess an integrated form of a related phage called P2. λ are therefore said to be Spi⁺ (sensitive to

(a) Insertional inactivation of the *lacZ'* gene

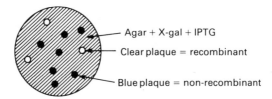

Agar + X-gal + IPTG

Clear plaque = recombinant

Blue plaque = non-recombinant

(b) Insertional inactivation of the λ*cI* gene

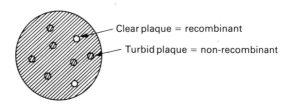

Clear plaque = recombinant

Turbid plaque = non-recombinant

(c) Selection using the Spi phenotype

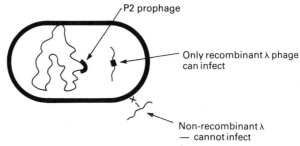

P2 prophage

Only recombinant λ phage can infect

Non-recombinant λ — cannot infect

(d) Selection on the basis of λ genome size

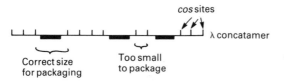

cos sites

λ concatamer

Correct size for packaging

Too small to package

Figure 5.12 Strategies for the selection of recombinant phage.

P2 prophage inhibition). Some λ cloning vectors are designed so that insertion of new DNA causes a change from Spi$^+$ to Spi$^-$, thus the recombinants can infect cells that carry P2 prophages. Such cells are used as the host for cloning experiments with these vectors; only recombinants are Spi$^-$ so only recombinants form plaques (Figure 5.12c).

5.4.4 SELECTION ON THE BASIS OF λ GENOME SIZE

The λ packaging system, which assembles the mature phage particles, can only insert DNA molecules of between 37 and 52 kb into the head structure. Anything less than 37 kb will not be packaged. Many λ vectors have been constructed by deleting large segments of the λ DNA molecule (p. 118) and so are less than 37 kb in length. These will only package into mature phage particles after extra DNA has been inserted, bringing the total genome size up to 37 kb or more (Figure 5.12d). Therefore with these vectors only recombinant phage are able to replicate.

5.5 TRANSFORMATION OF NON-BACTERIAL CELLS

Ways of introducing DNA into yeast, fungi, animals and plants are also needed if these organisms are to be used as the hosts for gene cloning. In general terms, soaking cells in salt is effective only with a few species of bacteria, although treatment with lithium chloride or lithium acetate does enhance DNA uptake by yeast cells, and is frequently used in the transformation of *Saccharomyces cerevisiae*. However, for most higher organisms more sophisticated methods are needed.

5.5.1 TRANSFORMATION OF INDIVIDUAL CELLS

With most organisms the main barrier to DNA uptake is the cell wall. Cultured animal cells, which usually lack cell walls, are easily transformed, especially if the DNA is precipitated on to the cell surface with calcium phosphate (Figure 5.13a). For other types of cell the answer is often to remove the cell wall. Enzymes that degrade yeast, fungal and plant cell walls are available, and under the right conditions intact **protoplasts** can be obtained (Figure 5.13b). Protoplasts generally take up DNA quite readily; alternatively transformation can be stimulated by special techniques such as **electroporation**, during which the cells are subjected to a short electrical pulse, thought to induce the transient formation of pores in the cell membrane through which the DNA molecules can enter the cell. After transformation the protoplasts are washed to remove the degradative enzymes and the cell wall spontaneously reforms.

In contrast to the transformation systems described so far there are two physical methods for introducing DNA into cells. The first of these is **microinjection**, which makes use of a very fine pipette to inject DNA

(a) Precipitation of DNA on to animal cells

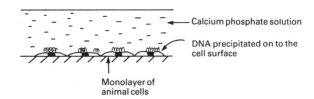

Calcium phosphate solution

DNA precipitated on to the cell surface

Monolayer of animal cells

(b) Transformation of plant protoplasts

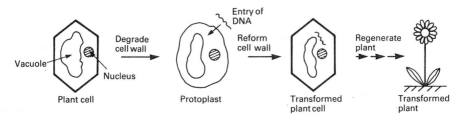

Vacuole

Nucleus

Plant cell

Degrade cell wall

Entry of DNA

Protoplast

Reform cell wall

Transformed plant cell

Regenerate plant

Transformed plant

Figure 5.13 Strategies for introducing new DNA into animal and plant cells.

molecules directly into the nucleus of the cells to be transformed (Figure 5.14a). This technique was initially applied to animal cells but has subsequently been successful with plant cells. The second method involves bombardment of the cells with high-velocity microprojectiles, usually particles of gold or tungsten that have been coated with DNA. These microprojectiles are fired at the cells from a particle gun (Figure 5.14b). This unusual technique is termed **biolistics** and has been used with a number of different types of cell.

5.5.2 TRANSFORMATION OF WHOLE ORGANISMS

With animals and plants the desired end-product may not be transformed cells, but a transformed organism. Plants are relatively easy to regenerate from cultured cells, though problems have been experienced in developing regeneration procedures for monocotyledonous species such as cereals and grasses. A single transformed plant cell can therefore give rise to a transformed plant, which will carry the cloned DNA in every cell, and will pass the cloned DNA on to its progeny following flowering and seed formation (Figure 7.13b). Animals of course cannot be regenerated from cultured cells, so obtaining transformed animals requires a rather more subtle approach. The standard technique with mammals such as mice is to remove fertilized eggs from the oviduct, to microinject DNA, and then to re-implant the transformed cells into the mother's reproductive tract.

(a) Microinjection

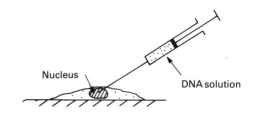

(b) Transformation with microprojectiles

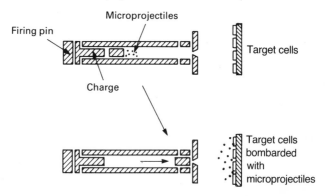

Figure 5.14 Two physical methods for introducing DNA into cells.

FURTHER READING

Cohen, S. N. *et al.* (1972) Nonchromosomal antibiotic resistance in bacteria: genetic transformation of *Escherichia coli* by R-factor DNA. *Proceedings of the National Academy of Sciences, USA*, **69**, 2110–14 – transformation of a bacterium with a plasmid.

Mandel, M. and Higa, A. (1970) Calcium-dependent bacteriophage DNA infection. *Journal of Molecular Biology*, **53**, 154–62 – transfection.

Hohn, B. and Murray, K. (1977) Packaging recombinant DNA molecules into bacteriophage particles *in vitro*. *Proceedings of the National Academy of Sciences, USA*, **74**, 3259–63 – *in vitro* packaging.

Capecchi, M. R. (1980) High efficiency transformation by direct microinjection of DNA into cultured mammalian cells. *Cell*, **22**, 479–88.

Klein, T. M. *et al.* (1987) High velocity microprojectiles for delivering nucleic acids into living cells. *Nature*, **327**, 70–3 – biolistics.

Cloning vectors for *E. coli*

6

The basic experimental techniques involved in gene cloning have now been described. In Chapters 3, 4 and 5 we have seen how DNA can be purified from cell extracts, how recombinant DNA molecules can be constructed in the test-tube, how DNA molecules can be reintroduced into living cells, and how recombinant clones can be distinguished. Now we must look more closely at the cloning vector itself, in order to consider the range of vectors available to the molecular biologist, and to understand the properties and uses of each individual type.

The greatest variety of cloning vectors exist for use with *E. coli* as the host organism. This is not surprising in view of the central role that this bacterium has played in basic research over the last 50 years. The tremendous wealth of information that exists concerning the microbiology, biochemistry and genetics of *E. coli* has meant that virtually all fundamental studies of gene structure and function have been carried out with this bacterium as the experimental organism. In recent years, gene cloning and molecular biological research have become mutually synergistic – breakthroughs in gene cloning have acted as a stimulus to research, and the needs of research have spurred on the development of new, more sophisticated cloning vectors.

In this chapter the most important types of *E. coli* cloning vector will be described, and the specific uses of representative molecules outlined. In Chapter 7, cloning vectors for yeast, fungi, plants and animals will be considered.

6.1 CLONING VECTORS BASED ON *E. coli* PLASMIDS

The simplest cloning vectors, and the ones in most widespread use in gene cloning, are those based on small bacterial plasmids. A large number of different plasmid vectors are available for use with *E. coli*, many obtainable from commercial suppliers. They combine ease of purification with desirable properties such as high transformation efficiency, convenient selectable markers for transformants and recombinants, and the ability to clone reasonably large (up to about 5 kb) pieces of DNA. Most 'routine' gene cloning experiments make use of one or other of these plasmid vectors.

Probably the most popular vector is pBR322, which was introduced in Chapter 5 to illustrate selection by insertional inactivation of antibiotic resistance genes (p. 90). We will now look more closely at this plasmid.

6.1.1 THE NOMENCLATURE OF PLASMID CLONING VECTORS

The name 'pBR322' conforms with the standard rules for vector nomenclature.

'p' indicates that this is indeed a plasmid

'BR' identifies the laboratory in which the vector was originally constructed (BR stands for Bolivar and Rodriguez, the two researchers who developed pBR322)

'322' distinguishes this plasmid from others developed in the same ·laboratory (there are also plasmids called pBR325, pBR327, pBR328, etc.)

6.1.2 THE USEFUL PROPERTIES OF pBR322

The genetic and physical map of pBR322 (Figure 6.1) gives an indication of why this plasmid has become such a popular cloning vector.

The first useful feature of pBR322 is its size. In Chapter 2 it was stated that a cloning vector ought to be less than 10 kb in size, to avoid problems such as DNA breakdown during purification. pBR322 is 4363 bp in size, which means that not only can the vector itself be purified with ease, but so can recombinant DNA molecules constructed with it. Even with 6 kb of additional DNA, a recombinant pBR322 molecule will still be a manageable size.

The second feature of pBR322 is that, as described in Chapter 5, it carries two sets of antibiotic resistance genes. Either ampicillin or tetracycline resistance can be used as a selectable marker for cells containing the plasmid, and each marker gene includes unique restriction sites that can be used in cloning experiments. Insertion of new DNA into pBR322 that has been restricted with *Pst*I, *Pvu*I or *Sca*I will inactivate the amp^R gene, and insertion using any one of eight restriction endonucleases (notably *Bam*HI and *Hind*III) inactivates tetracycline resistance. This great variety of restriction sites that can be used for insertional inactivation means that pBR322 can be used to clone DNA fragments with any of several kinds of sticky end.

A third advantage of pBR322 is that it has a reasonably high copy number. Generally there are about 15 molecules present in a transformed *E. coli* cell, but this number can be increased, up to 1000–3000, by plasmid amplification in the presence of a protein synthesis inhibitor such as chloramphenicol (p. 41). An *E. coli* culture will therefore provide a good yield of recombinant pBR322 molecules.

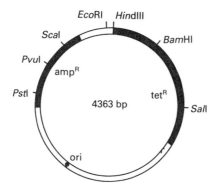

Figure 6.1 A map of pBR322 showing the positions of the ampicillin resistance (amp^R) and tetracycline resistance (tet^R) genes, the origin of replication (ori) and a selection of the most important restriction sites.

6.1.3 THE PEDIGREE OF pBR322

The remarkable convenience of pBR322 as a cloning vector did not arise by chance. The plasmid was in fact designed in such a way that the final construct would possess these desirable properties. An outline of the scheme used to construct pBR322 is shown in Figure 6.2a. It can be seen that its production was a tortuous business that required full and skilful use of the DNA manipulative techniques described in Chapter 4. A summary of the result of these manipulations is provided in Figure 6.2b, from which it can be seen that pBR322 in fact comprises DNA derived from three different naturally occurring plasmids. The amp^R gene originally resided on the plasmid R1, a typical antibiotic resistance plasmid that occurs in natural populations of *E. coli* (p. 17). The tet^R gene is derived from pSC101. The replication origin of pBR322, which directs multiplication of the vector in host cells, is originally from pMB1, which is closely related to the colicin-producing plasmid ColE1 (p. 17).

6.1.4 OTHER TYPICAL *E. coli* PLASMID CLONING VECTORS

It would be pointless to describe all the plasmid cloning vectors that are available for use with *E. coli*, especially as many of these are very similar to pBR322 in their properties and uses. However, four additional vectors should be mentioned. Each is derived from pBR322, but each displays an extra feature that, for some gene cloning experiments, makes it a better choice than pBR322.

(a) pBR325 – three selectable markers (Figure 6.3a)

pBR325 is basically pBR322 with an additional segment of DNA. This segment carries the gene for chloramphenicol acetyltransferase (CAT), an

(a) Construction of pBR322

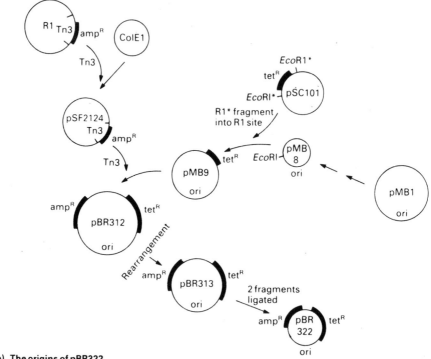

(b) The origins of pBR322

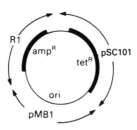

Figure 6.2 The pedigree of pBR322. (a) The manipulations involved in construction of pBR322. (b) A summary of the origins of pBR322.

enzyme that inactivates chloramphenicol and provides resistance to this antibiotic. The CAT gene contains the only *Eco*R1 restriction site on the plasmid, so this site can be used for cloning with recombinants detected by insertional inactivation of chloramphenicol resistance. pBR325 therefore has the advantage over pBR322 of an additional cloning site (pBR322 has an

(a) pBR325

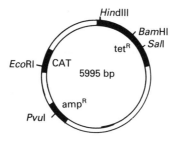

(b) pAT153 and pBR327

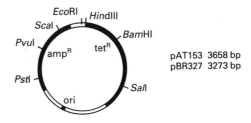

(c) pUC8

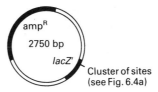

Figure 6.3 *E. coli* plasmid cloning vectors.

*Eco*R1 site but it does not lie in either the amp^R or tet^R genes) and an additional antibiotic resistance gene to use as a selectable marker.

(b) pAT153 and pBR327 – high copy number plasmids (Figure 6.3b)

These two vectors are both derivatives of pBR322, but the alterations do not involve addition of a new selectable marker or extra cloning sites. Instead, pAT153 and pBR327 were constructed by removing segments of pBR322, 705 bp in the case of pAT153 and 1089 bp for pBR327. These deletions leave the amp^R and tet^R genes intact, but change the replicative and conjugative abilities of the resulting plasmids. Each plasmid differs from pBR322 in two ways.

1. pAT153 and pBR327 have higher copy numbers than pBR322, being present at about 30–45 molecules per *E. coli* cell. This is not of great relevance as far as plasmid yield is concerned, as each of the three plasmids can be amplified to copy numbers greater than 1000. However, the higher copy numbers of pAT153 and pBR327 in normal cells make these vectors particularly useful if the aim of the experiment is to study the function of the cloned gene. In these cases gene dosage becomes important, as the more copies there are of a cloned gene, the more likely it is that the effect of the cloned gene on the host cell will be detectable. pAT153 and pBR327, with their higher copy numbers, are therefore more suitable than pBR322 for this kind of work.

2. The deletions destroy the conjugative ability of pBR322, making pAT153 and pBR327 non-conjugative plasmids that cannot direct their own transfer to other *E. coli* cells. This is important for **biological containment**, averting the possibility of a recombinant pAT153 or pBR327 molecule escaping from the test-tube and colonizing bacteria in the gut of a careless molecular biologist. In contrast, pBR322 could theoretically be passed to natural populations of *E. coli* by conjugation, though in fact pBR322 does itself carry safeguards to minimize the chances of this happening. pAT153 and pBR327 are, however, preferable if the cloned gene is potentially harmful should an accident occur.

(c) pUC8 – selection using the *lacZ'* gene (Figure 6.3c)

This vector was mentioned in Chapter 5 when selection of recombinants using insertional inactivation of the β-galactosidase gene was described (p. 93). pUC8 is in fact derived from pBR322, although in this case only the replication origin and the amp^R gene remain. The nucleotide sequence of this gene has been changed so that it no longer contains the unique restriction sites; all these cloning sites are now clustered into a short segment of the *lacZ'* gene carried by pUC8.

pUC8 has two important advantages that have resulted in it becoming a very popular cloning vector. The first is that identification of *E. coli* cells containing recombinant pUC8 molecules can be achieved by a single-step process, by plating on to agar medium containing ampicillin plus X-gal (p. 93). With each of the other vectors described so far, selection for recombinants is a two-step procedure, requiring replica-plating from one antibiotic medium to another, as described on p. 91 for pBR322. The second advantage of pUC8 lies with the clustering of the restriction sites, which allows a DNA fragment with two different sticky ends (say *Eco*R1 at one end and *Bam*HI at the other) to be cloned without resorting to additional manipulations such as linker attachment (Figure 6.4a). Other pUC vectors

(a) Restriction sites in pUC8

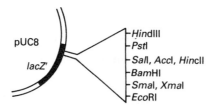

(b) Restriction sites in pUC10

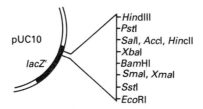

(c) Shuttling a DNA fragment from pUC8 to M13mp8

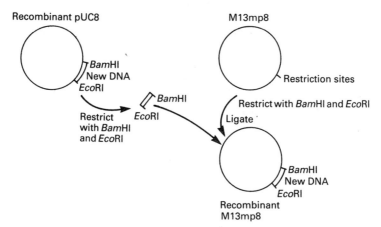

Figure 6.4 The pUC plasmids. (a) The restriction site cluster in the *lacZ'* gene of pUC8. (b) The restriction site cluster in pUC10. (c) Shuttling a DNA fragment from pUC8 to M13mp8.

carry different combinations of restriction sites and provide even greater flexibility in the types of DNA fragment that can be cloned (Figure 6.4b). Furthermore, the restriction site clusters in these vectors are the same as the clusters in the equivalent M13mp series of vectors (p. 112). DNA cloned into pUC8 can therefore be transferred directly to M13mp8 in which form it can be analysed by DNA sequencing or *in vitro* mutagenesis (Figure 6.4c; see p. 186 and 215 for descriptions of these procedures).

6.2 CLONING VECTORS BASED ON M13 BACTERIOPHAGE

The most essential requirement for any cloning vector is that it has a means of replicating in the host cell. For plasmid vectors this requirement is easy to satisfy, as relatively short DNA sequences are able to act as plasmid origins of replication, and most, if not all, of the enzymes needed for replication are provided by the host cell. Elaborate manipulations, such as those that resulted in pBR322 (Figure 6.2a), are therefore possible so long as the final construction has an intact, functional replication origin.

With bacteriophages such as M13 the situation as regards replication is more complex. Phage DNA molecules generally carry several genes that are essential for replication, including genes coding for components of the phage protein coat and phage-specific DNA replicative enzymes. Alteration or deletion of any of these genes will impair or destroy the replicative ability of the resulting molecule. There is therefore much less freedom to modify phage DNA molecules, and generally phage cloning vectors are only slightly different from the parent molecule.

The problems in constructing a phage cloning vector are illustrated by considering M13. The normal M13 genome is 6.4 kb in length, but most of this is taken up by ten closely packed genes (Figure 6.5), each essential for the replication of the phage. There is only a single, 507 nucleotide intergenic sequence into which new DNA could be inserted without disrupting one of

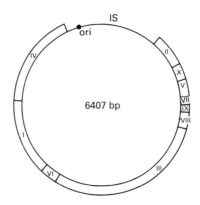

Figure 6.5 The M13 genome showing the positions of genes I to X.

these genes, and in fact this region includes the replication origin which must itself remain intact. Clearly there is only limited scope for modifying the M13 genome.

Nevertheless, it will be remembered that the great attraction of M13 is the opportunity it offers of obtaining single-stranded versions of cloned DNA (p. 24). This feature has acted as a stimulus for the development of M13 cloning vectors.

6.2.1 DEVELOPMENT OF THE CLONING VECTOR M13mp2

The first step in construction of an M13 cloning vector was to introduce the *lacZ'* gene into the intergenic sequence. This gave rise to M13mp1, which forms blue plaques on X-gal agar (Figure 6.6a).

M13mp1 does not possess any unique restriction sites in the *lacZ'* gene. It does, however, contain the hexanucleotide GGATTC, near the start of the gene. A single nucleotide change would make this GAATTC, which is an *Eco*RI site. This alteration was carried out using *in vitro* mutagenesis (p. 213), resulting in M13mp2 (Figure 6.6b). M13mp2 has a slightly altered *lacZ'* gene (the fifth codon now specifies asparagine instead of aspartic acid), but the β-galactosidase enzyme produced by cells infected with M13mp2 is still perfectly functional.

M13mp2 is the simplest M13 cloning vector. DNA fragments with *Eco*RI sticky ends can be inserted into the cloning site, and recombinants are distinguished as clear plaques on X-gal agar.

6.2.2 M13mp7 – SYMMETRICAL CLONING SITES

The next step in the development of M13 vectors was to introduce additional restriction sites into the *lacZ'* gene. This was achieved by synthesizing in the test-tube a short oligonucleotide, called a **polylinker**, that consists of a series of restriction sites and has *Eco*RI sticky ends (Figure 6.7a). This polylinker was inserted into the *Eco*RI site of M13mp2, to give M13mp7, a more complex vector with four possible cloning sites (*Eco*RI, *Bam*HI, *Sal*I and *Pst*I). The polylinker is designed so that it does not totally disrupt the *lacZ'* gene; a reading frame is maintained throughout the polylinker, and a functional, though altered, β-galactosidase enzyme is still produced (Figure 6.7b).

When M13mp7 is digested with either *Eco*RI, *Bam*HI or *Sal*I, a part or all of the polylinker is excised (Figure 6.8a). On ligation, in the presence of new DNA, one of three events may occur (Figure 6.8b).

1. New DNA is inserted.
2. The polylinker is reinserted.
3. The vector self-ligates without insertion.

Insertion of new DNA almost invariably prevents β-galactosidase production, so recombinant plaques are clear on X-gal agar (Figure 6.8c). Alternatively, if the polylinker is reinserted, and the original M13mp7

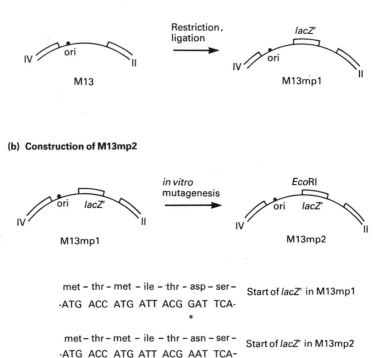

(a) Construction of M13mp1

Restriction, ligation

M13

M13mp1

(b) Construction of M13mp2

in vitro mutagenesis

M13mp1

M13mp2

met – thr – met – ile – thr – asp – ser –   Start of *lacZ'* in M13mp1
-ATG ACC ATG ATT ACG GAT TCA-
                              *

met – thr – met – ile – thr – asn – ser –   Start of *lacZ'* in M13mp2
-ATG ACC ATG ATT ACG AAT TCA-
                         *
                      *Eco*RI

Figure 6.6 Construction of (a) M13mp1 and (b) M13mp2 from the wild-type M13 genome.

reformed, then blue plaques will result. But what if the vector self-ligates, with neither new DNA nor the polylinker inserted? Once again the design of the polylinker comes into play. Whichever restriction site is used, self-ligation results in a functional *lacZ'* gene (Figure 6.8c), giving blue plaques. Selection is therefore unequivocal: only recombinant M13mp7 phage give rise to clear plaques.

The big advantage of M13mp7, with its symmetrical cloning sites, is that DNA inserted into either the *Bam*HI, *Sal*I or *Pst*I sites can be excised from the recombinant molecule using *Eco*RI (Figure 6.9). Very few vectors allow cloned DNA to be recovered so easily.

6.2.3 MORE COMPLEX M13 VECTORS

The latest M13 vectors have more complex polylinkers inserted into the *lacZ'* gene. An example is M13mp8 (Figure 6.10a), which is the counterpart of the

(a) The polylinker

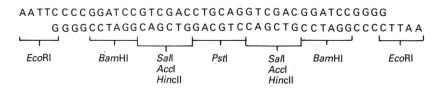

(b) Construction of M13mp7

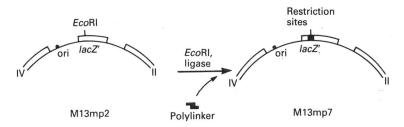

Figure 6.7 Construction of M13mp7. (a) The polylinker and (b) its insertion into the *Eco*RI site of M13mp2.

plasmid pUC8 (p. 108). As with the plasmid vector, one advantage of M13mp8 is its ability to take DNA fragments with two different sticky ends.

A second feature is provided by the sister vector M13mp9 (Figure 6.10b), which has the same polylinker but in the reverse orientation. A DNA fragment cloned into M13mp8, if excised by double restriction, and then inserted into M13mp9, will now itself be in the reverse orientation (Figure 6.10c). This is important in DNA sequencing (p. 186), in which the nucleotide sequence is read from one end of the polylinker into the inserted DNA fragment (Figure 6.10d). Only about 400 nucleotides can be read from one sequencing experiment; if the inserted DNA is longer than this, one end of the fragment will not be sequenced. The answer is to turn the fragment around, by excising and reinserting into the sister vector. A DNA sequencing experiment with this new clone will allow the nucleotide sequence at the other end of the fragment to be determined.

Other M13 vector pairs are also available. M13mp10/11 and M13mp18/19 are similar to M13mp8/9, but have different polylinkers and therefore different restriction sites.

6.2.4 HYBRID PLASMID–M13 VECTORS

Although M13 vectors are very useful for the production of single-stranded versions of cloned genes they do suffer from one disadvantage. There is a limit to the size of DNA fragment that can be cloned with an M13 vector,

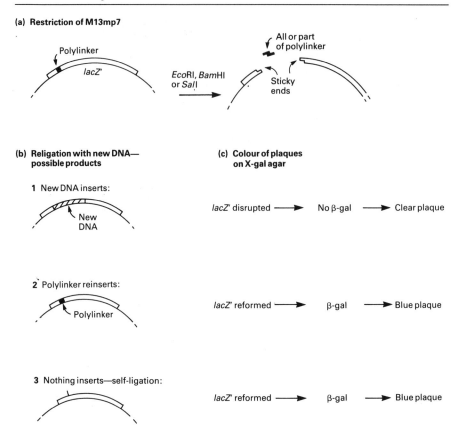

Figure 6.8 Cloning with M13mp7. See the text for details.

with 1500 bp generally being looked on as the maximum capacity, though fragments up to 3 kb have occasionally been cloned. To get around this problem a number of novel vectors have been developed by combining a part of the M13 genome with plasmid DNA. An example is provided by pEMBL8 (Figure 6.11a), which was made by transferring into pUC8 a 1300 bp fragment of the M13 genome. This piece of M13 DNA contains the signal sequence recognized by the enzymes that convert the normal double-stranded M13 molecule into single-stranded DNA before secretion of new phage particles. This signal sequence is still functional even though detached from the rest of the M13 genome, so pEMBL8 molecules are also converted into single-stranded DNA and will be secreted as defective phage particles (Figure 6.11b). All that is necessary is that the *E. coli* cells used as hosts for a pEMBL8 cloning experiment are subsequently infected with normal M13 to act as a **helper phage**, providing the necessary replicative enzymes and phage coat proteins. pEMBL8, being derived from pUC8, has the polylinker cloning sites within the *lacZ'* gene, so recombinant plaques

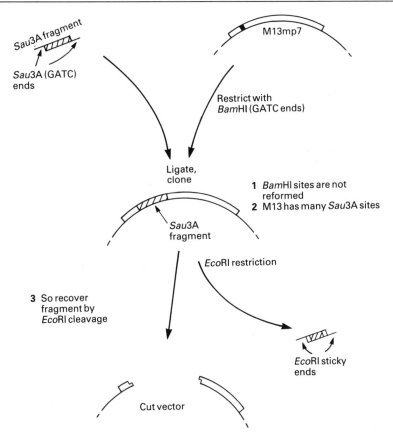

Figure 6.9 Recovery of cloned DNA from a recombinant M13mp7 molecule by restriction at the outer sites of the polylinker.

can be identified in the standard way on agar containing X-gal. With pEMBL8, single-stranded versions of cloned DNA fragments up to 10 kb in length can be obtained, greatly extending the range of the M13 cloning system.

6.3 CLONING VECTORS BASED ON λ BACTERIOPHAGE

Two problems had to be solved before λ-based cloning vectors could be developed:

1. The λ DNA molecule can be increased in size by only about 5%, representing the addition of only 3 kb of new DNA. If the total size of the molecule is more than 52 kb, then it will not package into the λ head structure and infective phage particles will not be formed. This severely

(a) The M13mp8/9 polylinker

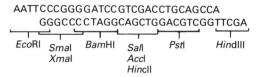

(b) The orientation of the polylinker

(c) Shuttling DNA from M13mp8 to M13mp9

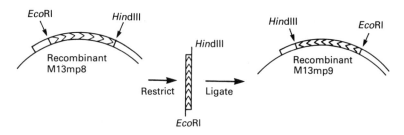

(d) DNA sequencing using M13mp8 and M13mp9

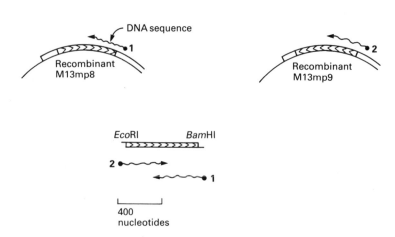

Figure 6.10 M13mp8 and M13mp9.

(a) pEMBL8

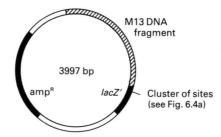

(b) Conversion of pEMBL8 into single-stranded DNA

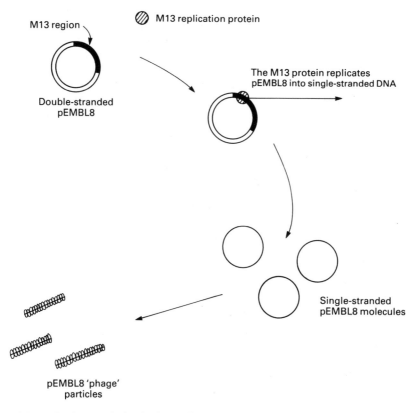

Figure 6.11 pEMBL8: a hybrid plasmid–M13 vector that can be converted into single-stranded DNA.

limits the size of a DNA fragment that can be inserted into an unmodified λ vector (Figure 6.12a).

2. The λ genome is so large that it has more than one recognition sequence for virtually every restriction endonuclease. Restriction cannot be used to cleave the normal λ molecule in a way that will allow insertion of new DNA, as the molecule would be cut into several small fragments that would be very unlikely to reform a viable λ genome on religation (Figure 6.12b).

In view of these difficulties it is perhaps surprising that a wide variety of λ cloning vectors have been developed, their primary use being to clone large pieces of DNA, from 5 kb to 40 kb, much too big to be handled by plasmid or M13 vectors.

6.3.1 SEGMENTS OF THE λ GENOME CAN BE DELETED WITHOUT IMPAIRING VIABILITY

The way forward for the development of λ cloning vectors was provided by the discovery that a large segment in the central region of the λ DNA molecule can be removed without affecting the ability of the phage to infect *E. coli* cells. Removal of all, or part, of this non-essential region, between positions 20 and 35 on the map shown in Figure 2.9, decreases the size of the resulting λ molecule by up to 15 kb. This means that as much as 18 kb of new DNA can now be added before the cut-off point for packaging is reached (Figure 6.13).

The 'non-essential' region in fact contains most of the genes involved in integration and excision of the λ prophage from the *E. coli* chromosome. A deleted λ genome is therefore non-lysogenic and can follow only the lytic infection cycle. This in itself is desirable for a cloning vector as it means induction is not needed before plaques are formed (p. 42).

6.3.2 NATURAL SELECTION CAN BE USED TO ISOLATE MODIFIED λ THAT LACK CERTAIN RESTRICTION SITES

Even a deleted λ genome, with the non-essential region removed, has multiple recognition sites for most restriction endonucleases. This is a problem that is often encountered when a new vector is being developed. If just one or two sites need to be removed then the technique of *in vitro* mutagenesis (p. 215) can be used. For example, an *Eco*RI site, GAATTC, could be changed to GGATTC, which is not recognized by the enzyme. However, *in vitro* mutagenesis was in its infancy when the first λ vectors were under development, and even today would not be an efficient means of changing more than a few sites in a single molecule.

Instead, natural selection was used to provide strains of λ that lack the unwanted restriction sites. Natural selection can be brought into play by using as a host an *E. coli* strain that produces *Eco*RI. Most λ DNA molecules that invade the cell will be destroyed by the restriction endonuclease;

(a) The size limitation

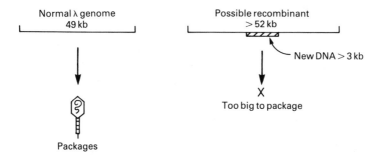

(b) Multiple restriction sites

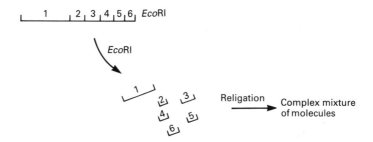

Figure 6.12 The two problems that had to be solved before λ cloning vectors could be developed. (a) The size limitation placed on the λ genome by the need to package it into the phage head. (b) λ DNA has multiple recognition sites for almost all restriction endonucleases.

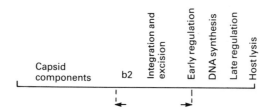

Figure 6.13 The λ genetic map, showing the position of the non-essential region that can be deleted without affecting the ability of the phage to follow the lytic infection cycle.

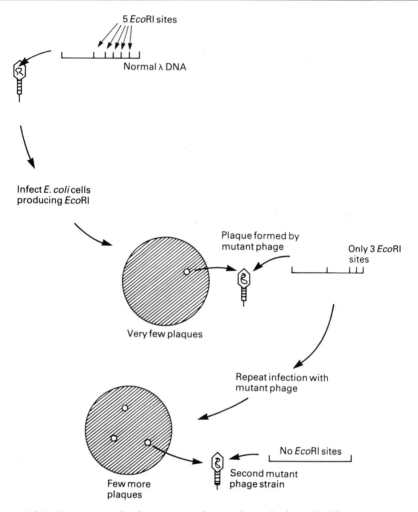

Figure 6.14 Using natural selection to isolate λ phage lacking *Eco*RI restriction sites.

however, a few will survive and produce plaques. These will be mutant phage, from which one or more *Eco*RI sites have been lost spontaneously (Figure 6.14). Several cycles of infection will eventually result in λ molecules that lack all or most of the *Eco*RI sites.

6.3.3 INSERTION AND REPLACEMENT VECTORS

Once the problems posed by packaging constraints and by the multiple restriction sites had been solved, the way was open for the development of different types of λ-based cloning vectors. The first two classes of vector to be produced were λ-**insertion** and λ-**replacement** (or substitution) vectors.

(a) Construction of a λ-insertion vector

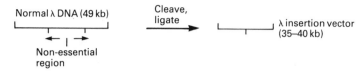

(b) λ gt10

(c) λ Charon 16

Figure 6.15 λ insertion vectors.

(a) Insertion vectors

With an insertion vector (Figure 6.15a), a large segment of the non-essential region has been deleted, and the two arms ligated together. An insertion vector will possess at least one unique restriction site into which new DNA can be inserted. The size of the DNA fragment that an individual vector can carry depends of course on the extent to which the non-essential region has been deleted. Two popular insertion vectors are:

λgt10 (Figure 6.15b), which can carry up to 8 kb of new DNA, inserted into a unique EcoRI site located in the cI gene. Insertional inactivation of this gene means that recombinants are distinguished as clear rather than turbid plaques (see p. 98).

λCharon 16 (Figure 6.15c), with which insertion of new DNA into a unique EcoRI site inactivates the lacZ' gene carried by the vector. Recombinants give clear, rather than blue, plaques on X-gal agar.

(b) Replacement vectors

A λ replacement vector has two recognition sites for the restriction endonuclease used for cloning. These sites flank a segment of DNA that is replaced by the DNA to be cloned (Figure 6.16a). Often the replaceable fragment (or 'stuffer fragment' in cloning jargon) carries additional restriction sites that can be used to cut it up into small pieces, so that its own reinsertion during a cloning experiment is very unlikely. Replacement vectors are generally designed to carry larger pieces of DNA than insertion vectors can handle. Recombinant selection is often on the basis of size, with non-recombinant vectors being too small to be packaged into λ phage heads (p. 100).

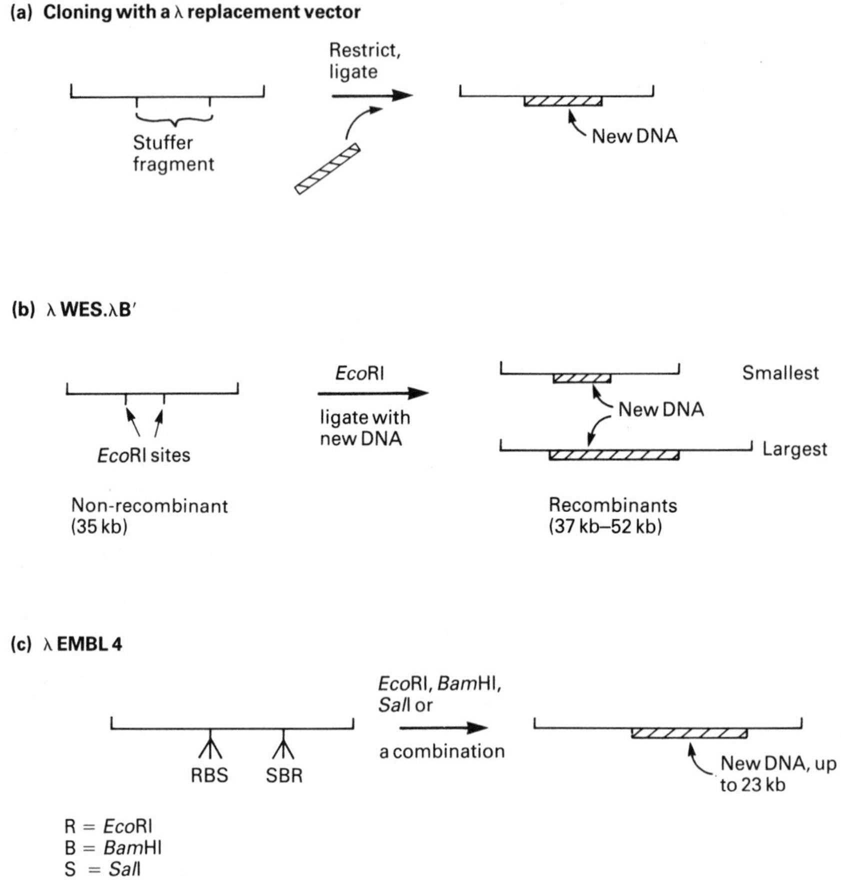

(a) Cloning with a λ replacement vector

(b) λ WES.λB′

(c) λ EMBL 4

R = *Eco*RI
B = *Bam*HI
S = *Sal*I

Figure 6.16 λ replacement vectors.

The most popular replacement vectors are:

λ**WES.λB′** (Figure 6.16b), in which two *Eco*RI sites flank the replacement fragment, and recombinant selection is solely on the basis of size. Inserts of up to 15 kb can be cloned.

λ**EMBL4** (Figure 6.16c), which can carry up to 23 kb of inserted DNA (near the theoretical maximum) by replacing a segment flanked by pairs of *Eco*RI, *Bam*HI and *Sal*I sites (either restriction endonuclease can be used to remove the stuffer fragment, so DNA fragments with a variety of sticky ends can be cloned). Recombinant selection with λEMBL4 can be on the basis of size, or can utilize the Spi phenotype (p. 98).

6.3.4 CLONING EXPERIMENTS WITH λ INSERTION OR REPLACEMENT VECTORS

A cloning experiment with a λ vector can proceed along the same lines as with a plasmid vector – the λ molecules are restricted, new DNA is added, the mixture is ligated, and the resulting molecules used to transfect a competent *E. coli* host (Figure 6.17a). This type of experiment requires that the vector be in its circular form, with the *cos* sites hydrogen-bonded to each other.

Although quite satisfactory for many purposes, a procedure based on transfection is not particularly efficient. A greater number of recombinants will be obtained if one or two refinements are introduced. The first is to purify the two arms of the vector. When the linear form of the vector is digested with the relevant restriction endonuclease, two fragments are produced, a left arm and a right arm (Figure 6.17b). With virtually all λ vectors the left arm is longer than the right, and the two can be separated and purified by sucrose density gradient centrifugation. A recombinant molecule is then constructed by mixing together the DNA to be cloned with samples of the left and right arm preparations. Ligation will result in several molecular arrangements, including concatamers comprising left arm–DNA–right arm repeated many times (Figure 6.17b). If the inserted DNA is the correct size then the *cos* sites that separate these structures will be the right distance apart for *in vitro* packaging (p. 95). Recombinant phage are therefore produced in the test-tube and can be used to infect an *E. coli* culture. This strategy, in particular the use of *in vitro* packaging, results in a large number of recombinant plaques.

6.3.5 VERY LARGE DNA FRAGMENTS CAN BE CLONED USING A COSMID

The final and most sophisticated type of λ-based vector is the **cosmid**. Cosmids are hybrids between a phage DNA molecule and a bacterial plasmid, and are designed around the fact that the enzymes that package the λ DNA molecule into the phage protein coat need only the *cos* sites in order to function (p. 24). The *in vitro* packaging reaction will work not only with λ genomes, but also with any molecule that carries *cos* sites separated by 37–52 kb of DNA.

A cosmid is basically a plasmid that carries a *cos* site (Figure 6.18a). It will also need a selectable marker, such as the ampicillin resistance gene, and a plasmid origin of replication, as cosmids lack all the λ genes and so do not produce plaques. Instead colonies are formed on selective media, just as with a plasmid vector.

A cloning experiment with a cosmid is carried out as follows (Figure 6.18b). The cosmid is opened at its unique restriction site and new DNA fragments inserted. These fragments are produced usually by partial digestion with a restriction endonuclease, as total digestion will invariably

(a) **Cloning with circular λ DNA**

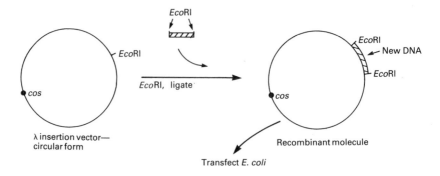

(b) **Cloning with linear λ DNA**

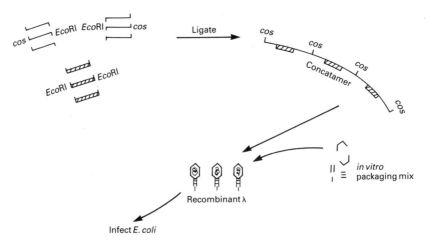

Figure 6.17 Different strategies for cloning with a λ vector. (a) Using the circular form of λ as a plasmid. (b) Using purified left and right arms of the λ genome, plus *in vitro* packaging, to achieve a greater number of recombinant plaques.

produce fragments too small to be cloned with a cosmid. Ligation is carried out so that concatamers are formed. Providing the inserted DNA is the right size, *in vitro* packaging will cleave the *cos* sites and place the recombinant cosmids in mature phage particles. These λ phage are then used to infect an *E. coli* culture, though of course plaques are not formed. Instead, infected cells are plated on to a selective medium and antibiotic-resistant colonies are grown. All colonies are recombinants as non-recombinant linear cosmids are too small to be packaged into λ heads.

(a) A typical cosmid

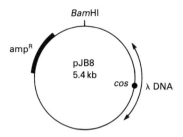

(b) Cloning with pJB8

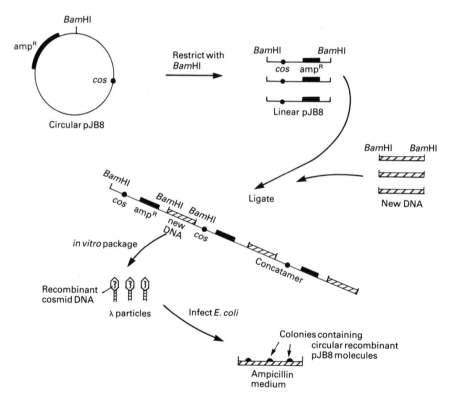

Figure 6.18 A typical cosmid and the way it is used to clone large fragments of DNA.

6.3.6 λ VECTORS ALLOW GENOMIC LIBRARIES TO BE CONSTRUCTED

The main use of all λ-based vectors is to carry DNA fragments that are too large to be handled by plasmid or M13 vectors. An insertion vector such as λgt10 can carry up to 8 kb of new DNA, and the λEMBL replacement vectors can manage 23 kb fragments. This compares with a maximum insert size of about 5 kb for most plasmids and less than 3 kb for M13 vectors.

But cosmids are used to clone the largest DNA fragments of all, up to 40 kb in some cases. This is because the cosmid itself can be quite small – just 8 kb or less is quite sufficient for the *cos* sites, selectable marker and plasmid replication origin. Cosmids therefore have a tremendous capacity for inserted DNA before the packaging cut-off point of about 52 kb is reached.

The ability to clone such large DNA fragments raises the possibility of the **genomic library**. A genomic library is a set of recombinant clones that contain all of the DNA present in an individual organism. An *E. coli* genomic library, for example, contains all the *E. coli* genes, so any desired gene can be withdrawn from the library and studied.

The big question is, how many clones will be needed for a genomic library. The answer can be calculated with the formula:

$$N = \frac{\ln (1 - P)}{\ln (1 - a/b)}$$

where N = number of clones required

P = probability (e.g. a 95% probability that any given gene is present)

a = average size of the DNA fragment inserted into the vector

b = total size of the genome

Table 6.1 shows the number of clones needed for genomic libraries of a variety of organisms. Although these figures may seem high they are by no means unreasonable. Bacterial, yeast, fungal and plant genomic libraries are now commonplace, and a few animal libraries also exist. These libraries can be retained for many years, and propagated so that copies can be sent from research group to research group. Their development, by providing depositories of cloned material, has greatly aided the progress of molecular biological and genetic research.

6.4 VECTORS FOR OTHER BACTERIA

Cloning vectors have also been developed for several other species of bacteria, including *Streptomyces*, *Bacillus* and *Pseudomonas*. Some of these vectors are based on plasmids specific to the host organism, and some on **broad host range plasmids** that will replicate in a variety of bacterial hosts. A few are derived from bacteriophages specific to these organisms; for example, several *Streptomyces* cloning vectors are based on a λ-like phage

Table 6.1 Number of clones needed for genomic libraries of a variety of organisms

Species	Genome size (bp)		Number of clones[a]	
			17 kb fragments[b]	35 kb fragments[c]
Escherichia coli	4	$\times\ 10^6$	700	340
Bacillus megaterium	3	$\times\ 10^7$	5 300	2 600
Aspergillus nidulans	4	$\times\ 10^7$	7 000	3 400
Saccharomyces cerevisiae	7	$\times\ 10^7$	12 500	6 000
Drosophila melanogaster	8	$\times\ 10^7$	14 100	6 850
Tomato	7	$\times\ 10^8$	125 000	60 000
Man	3	$\times\ 10^9$	535 000	258 250
Frog	2.3	$\times\ 10^{10}$	4 280 000	1 997 000

[a] Calculated for a probability (P) of 95% that any particular gene will be present in the library.
[b] Fragments suitable for a replacement vector such as λEMBL4.
[c] Fragments suitable for a cosmid.

called φC31. Most of these vectors are very similar to E. coli vehicles in terms of general purposes and uses and need not be considered in this book.

FURTHER READING

Pouwels, P. H. et al. (1985) Cloning Vectors. Elsevier, Amsterdam – details of all cloning vectors.

Bolivar, F. et al. (1977) Construction and characterisation of new cloning vectors. II. A multi-purpose cloning system. Gene, 2, 95–113 – pBR322.

Sanger, F. et al. (1980) Cloning in single-stranded bacteriophage as an aid to rapid DNA sequencing. Journal of Molecular Biology, 143, 161–78 – M13 vectors.

Dente, L. et al. (1983) pEMBL: a new family of single-stranded plasmids. Nucleic Acids Research, 11, 1645–55.

Leder, P. et al. (1977) EK2 derivatives of bacteriophage lambda useful in cloning of DNA from higher organisms: the gtWES system. Science, 196, 175–7.

Frischauf, A.-M. et al. (1983) Lambda replacement vectors carrying polylinker sequences. Journal of Molecular Biology, 170, 827–42 – the λEMBL vectors.

Collins, J. and Hohn, B. (1978) Cosmids: a type of plasmid gene cloning vector that is packageable in vitro in bacteriophage heads. Proceedings of the National Academy of Sciences, USA, 75, 4242–6.

Cloning vectors for organisms other than *E. coli* 7

Most cloning experiments are carried out with *E. coli* as the host, and the widest variety of cloning vectors are available for this organism. *E. coli* is particularly popular when the aim of the cloning experiment is to study the basic features of molecular biology such as gene structure and function. However, under some circumstances it may be desirable to use a different host for a gene cloning experiment. This is especially true in biotechnology, where the aim may not be to study a gene, but to use cloning to control or improve synthesis of an important metabolic product (for example, a hormone such as insulin), or to change the properties of the organism (for example to introduce herbicide resistance into a crop plant). We must therefore consider cloning vectors for organisms other than *E. coli*.

7.1 VECTORS FOR YEAST AND OTHER FUNGI

The yeast *Saccharomyces cerevisiae* is one of the most important organisms in biotechnology. As well as its role in brewing and breadmaking, yeast has been used as a host organism for the production of important pharmaceuticals from cloned genes (p. 240). Development of cloning vectors for yeast has been stimulated greatly by the discovery of a plasmid that is present in most strains of *S. cerevisiae* (Figure 7.1). The 2 μm circle, as it is called, is one of only a very limited number of plasmids found in eukaryotic cells.

7.1.1 SELECTABLE MARKERS FOR THE 2 μm PLASMID

The 2 μm circle is in fact an excellent basis for a cloning vector. It is 6 kb in size, which is ideal for a vector, and exists in the yeast cell at a copy number of between 70 and 200. Replication makes use of a plasmid origin, several enzymes provided by the host cell, and the proteins coded by the *REP1* and *REP2* genes carried by the plasmid.

However, all is not perfectly straightforward in using the 2 μm plasmid as a cloning vector. Firstly, there is the question of a selectable marker. One or

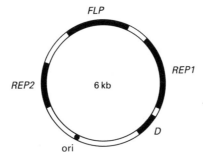

Figure 7.1 The yeast 2 μm circle. *REP1* and *REP2* are involved in replication of the plasmid, and *FLP* codes for a protein that can convert the A form of the plasmid (shown here) to the B form, in which the gene order has been rearranged by intramolecular recombination. The function of *D* is not exactly known.

two of the most recent yeast cloning vectors carry genes conferring resistance to inhibitors such as methotrexate and copper, but most of the popular yeast vectors make use of a radically different type of selection system. In practice a normal yeast gene is used, generally one that codes for an enzyme involved in amino acid biosynthesis. An example is the gene *LEU2*, which codes for β-isopropyl malate dehydrogenase, one of the enzymes involved in the conversion of pyruvic acid to leucine.

In order to use *LEU2* as a selectable marker, a special kind of host organism is needed. The host must be an **auxotrophic** mutant that has a non-functional *LEU2* gene. Such a *leu2⁻* yeast will be unable to synthesize leucine and will normally survive only if this amino acid is supplied as a nutrient in the growth medium (Figure 7.2a). Selection is possible because transformants will contain a plasmid-borne copy of the *LEU2* gene, and so will be able to grow in the absence of the amino acid. In a cloning experiment, cells are plated out on to **minimal medium** (which contains no added amino acids). Only transformed cells will be able to survive and form colonies (Figure 7.2b).

7.1.2 VECTORS BASED ON THE 2 μm CIRCLE – YEAST EPISOMAL PLASMIDS

Vectors derived from the 2 μm circle are called **yeast episomal plasmids** or YEps. A typical example of a YEp is provided by pJDB219 (Figure 7.3).

pJDB219 illustrates several general features of yeast cloning vectors. Firstly, it is a **shuttle vector**. As well as the 2 μm circle and the selectable *LEU2* gene, pJDB219 also includes the entire pBR322 sequence, and can therefore replicate and be selected for in both yeast and *E. coli*. There are several lines of reasoning behind the use of shuttle vectors. One is that it may be difficult to recover the recombinant DNA molecule from a transformed yeast colony. This is not such a problem with YEps, which are

(a) *LEU2* yeast

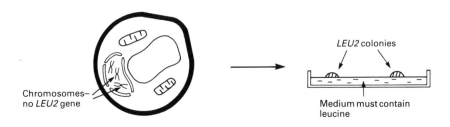

Chromosomes–
no *LEU2* gene

LEU2 colonies

Medium must contain
leucine

(b) Using *LEU2* as a selectable marker

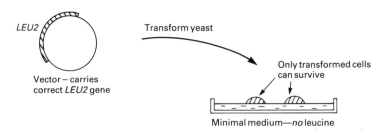

LEU2

Vector – carries
correct *LEU2* gene

Transform yeast

Only transformed cells
can survive

Minimal medium—*no* leucine

Figure 7.2 Using the *LEU2* gene as a selectable marker in a yeast cloning experiment.

present in yeast cells primarily as plasmids, but with other yeast vectors, which may integrate into one of the yeast chromosomes (see section 7.1.4), purification may be impossible. This is a disadvantage as in many cloning experiments purification of recombinant DNA is essential in order for the correct construct to be identified by, for example, DNA sequencing.

The standard procedure when cloning in yeast is therefore to perform the initial cloning experiment with *E. coli*, and to select recombinants in this organism. Recombinant plasmids can then be purified, characterized, and the correct molecule introduced into yeast (Figure 7.4).

7.1.3 A YEp MAY INSERT INTO YEAST CHROMOSOMAL DNA

The word 'episomal' indicates that a YEp can replicate as an independent plasmid, but also implies that integration into one of the yeast chromosomes can occur (see the definition of 'episome' on p. 13). Integration occurs because the gene carried on the vector as a selectable marker is very closely homologous to the mutant version of the gene present in the yeast chromosomal DNA. With pJDB219, for example, recombination can occur

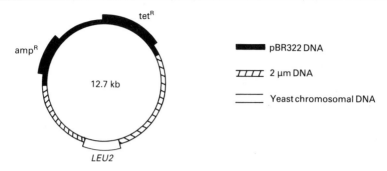

Figure 7.3 A yeast episomal plasmid (YEp)–pJDB219.

between the plasmid *LEU2* gene and the yeast mutant *LEU2* gene, resulting in insertion of the entire plasmid into one of the yeast chromosomes (Figure 7.5). The plasmid may remain integrated, or a later recombination event may result in it being excised again.

7.1.4 OTHER TYPES OF YEAST CLONING VECTOR

In addition to YEps, there are several other types of cloning vector for use with *S. cerevisiae*. Two important ones are as follows.

1. **Yeast integrative plasmids (YIps)**, which are basically bacterial plasmids carrying a yeast gene. An example is YIp5, which is pBR322 with an inserted *URA3* gene (Figure 7.6a). This gene codes for orotidine-5′-phosphate decarboxylase (an enzyme that catalyses one of the steps in the biosynthesis pathway for pyrimidine nucleotides) and is used as a selectable marker in exactly the same way as *LEU2*. A YIp will not replicate as a plasmid as it does not contain any parts of the 2 μm circle, but depends for its survival on integration into yeast chromosomal DNA. Integration occurs just as described for a YEp (Figure 7.5).
2. **Yeast replicative plasmids (YRps)**, which are able to multiply as independent plasmids because they carry a chromosomal DNA sequence that includes an origin of replication. Replication origins are known to be located very close to several yeast genes, including one or two which can be used as selectable markers. YRp7 (Figure 7.6b) is an example of a replicative plasmid. It is made up of pBR322 plus the yeast gene *TRP1*. This gene, which is involved in tryptophan biosynthesis, is located adjacent to a chromosomal origin of replication. The yeast DNA fragment present in YRp7 contains both *TRP1* and the origin.

Two factors come into play when deciding which type of yeast vector is most suitable for a particular cloning experiment. The first is **transformation frequency**. YEps have the highest transformation frequency, providing between 1000 and 100 000 transformed cells per μg. In contrast a YIp will

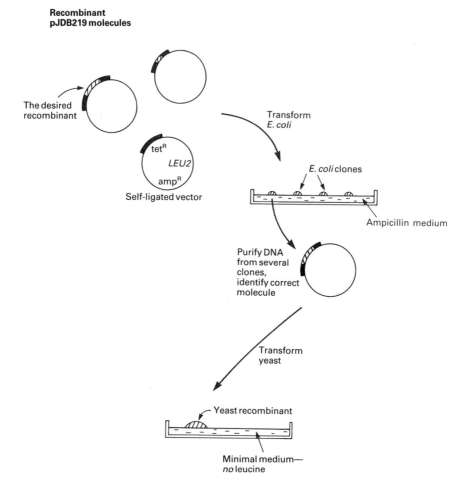

Figure 7.4 Cloning with an *E. coli*–yeast shuttle vector such as pJDB219.

yield only one to ten transformants per µg, a reflection of the fact that the rather rare chromosomal integration event is necessary before a YIp can be retained in a yeast cell.

The second factor to consider is the stability of the transformants. YEp transformants are not particularly stable; the plasmid may be lost spontaneously from a yeast cell and may also undergo molecular rearrangement, especially if a normal 2 µm plasmid is also present in the same cell. YRp transformants are also unstable. On the other hand, YIps produce extremely stable transformants, as loss of a YIp that has become integrated into chromosomal DNA is very unlikely.

Choice of a suitable vector therefore depends on the needs of the

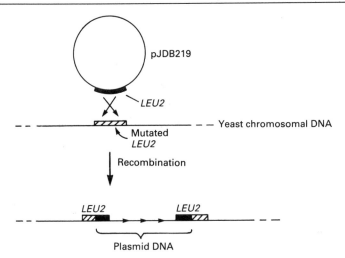

Figure 7.5 Recombination between plasmid and chromosomal *LEU2* genes can integrate pJDB219 into yeast chromosomal DNA. After integration there will be two copies of the *LEU2* gene; usually one will be functional, and the other mutated.

experiment. If a lot of transformants are required then a YEp or YRp should be used, but if stability is of paramount importance then a YIp will be more suitable.

7.1.5 ARTIFICIAL CHROMOSOMES CAN BE USED TO CLONE HUGE PIECES OF DNA IN YEAST

The final type of yeast cloning vector to consider is the **YAC**, which stands for **yeast artificial chromosome**, a totally new approach to gene cloning. Their development has been a spin-off from fundamental research into the structure of eukaryotic chromosomes, work that has identified the key components of a chromosome as being (Figure 7.7):

1. The centromere, which is required for the chromosome to be distributed correctly to daughter cells during cell division.
2. Two telomeres, the structures at the ends of a chromosome, which are needed in order for the ends to be replicated correctly and which also prevent the chromosome from being nibbled away by exonucleases.
3. The origins of replication, which are the positions along the chromosome at which DNA replication initiates, similar to the origin of replication of a plasmid (p. 12).

Once chromosome structure had been defined in this way the possibility arose that the individual components might be isolated by recombinant DNA techniques and then joined together again in the test-tube, creating an

(a) Ylp5

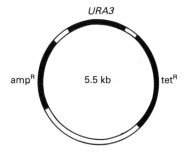

(b) YRp7

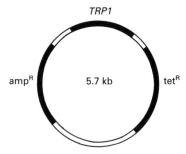

Figure 7.6 A Ylp and a YRp.

artificial chromosome. As the DNA molecules present in natural yeast chromosomes are several hundred kb in length, it should be possible with an artificial chromosome to clone pieces of DNA much larger than can be carried by any other type of vector.

(a) The structure and use of a YAC vector

Several YAC vectors have now been developed but each one is constructed along the same lines, with pYAC2 (Figure 7.8a) being a typical example. At first glance pYAC2 does not look much like an artificial chromosome, but on closer examination its unique features become apparent. pYAC2 is essentially a pBR322 plasmid into which a number of yeast genes have been inserted. Two of these genes, *URA3* and *TRP1*, have been encountered already as the selectable markers for YIp5 and YRp7 respectively. As in YRp7, the DNA fragment that carries *TRP1* also contains an origin of replication, but in pYAC2 this fragment is extended even further to include the sequence called *CEN4*, which is the DNA from the centromere region of

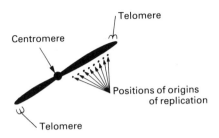

Figure 7.7 Chromosome structure.

chromosome 4. The *TRP1–origin–CEN4* fragment therefore contains two of the three components of the artificial chromosome.

The third component, the telomeres, are provided by the two sequences called *TEL*. These are not themselves complete telomere sequences, but once inside the yeast nucleus they will act as seeding sequences on to which telomeres will be built. This just leaves one other part of pYAC2 that has not been mentioned: *SUP4*, which is in fact the selectable marker into which new DNA is inserted during the cloning experiment.

The cloning strategy with pYAC2 is as follows (Figure 7.8b). The vector is first restricted with a combination of *Bam*HI and *Sma*I, cutting the molecule into three fragments. The *Bam*HI fragment is removed, leaving two arms, each bounded by one *TEL* sequence and one *Sma*I site. The DNA to be cloned, which must have blunt ends, is ligated between the two arms, producing the artificial chromosome. Protoplast transformation (p. 100) is then used to introduce the artificial chromosome into *S. cerevisiae*. The yeast strain that is used is a double auxotrophic mutant, *trp1⁻ ura3⁻*, which will be converted to *trp1⁺ ura3⁺* by the two markers on the artificial chromosome. Transformants are therefore selected by plating on to minimal medium, on which only cells containing a correctly constructed artificial chromosome will be able to grow. Any cells transformed with an incorrect artificial chromosome, containing two left or two right arms rather than one of each, will not be able to grow on minimal medium as one of the markers will be absent. The presence of the insert DNA in the vector can be checked by testing for insertional inactivation of *SUP4*, which is carried out by a simple colour test: white colonies are recombinants, red colonies are not.

(b) Applications for YAC vectors

The initial stimulus in designing artificial chromosomes came from yeast geneticists who wanted to use them to study various aspects of chromosome structure and behaviour, for instance to examine the segregation of chromosomes during meiosis. The applications of YACs in gene cloning experiments have only recently been explored. An important use of these vectors will be in obtaining intact clones of very long genes. Cosmids are the bacterial vectors with the largest capacity, but even these cannot handle

(a) pYAC2

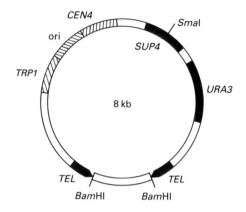

(b) The cloning strategy with pYAC2

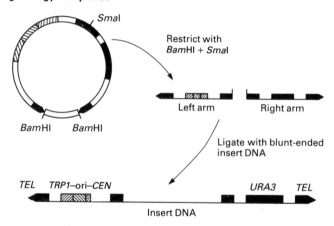

Figure 7.8 A YAC vector and the way it is used to clone large pieces of DNA.

pieces of DNA more than about 40 kb in length. Many important animal genes are much longer than this (the human cystic fibrosis gene, for instance, is 240 kb), and sometimes it is desirable to isolate gene clusters even longer than this. With a cosmid vector these genes and clusters cannot be cloned intact, and instead can only be obtained in several pieces, which is far from ideal. YACs offer a way round this problem and will allow the necessary fragments of DNA to be obtained.

More important though is the use of YAC vectors to produce gene libraries. Recall that a human gene library in a cosmid vector will comprise over a quarter million clones (Table 6.1). With a YAC vector this number can be reduced immensely, to just 60 000 clones if 150 kb fragments are used, making a human gene library much easier to work with.

7.1.6 VECTORS FOR OTHER YEAST AND FUNGI

Cloning vectors for other species of yeast and fungi are now available, and are helping in basic molecular biological analysis of these organisms, as well as extending the possible uses of yeasts and fungi in biotechnology. Episomal plasmids based on the *S. cerevisiae* 2 μm circle function well in related yeasts such as *Schizosaccharomyces pombe*, but do not appear to replicate in filamentous fungi such as *Aspergillus nidulans* and *Neurospora crassa*. Integrative plasmids, equivalent to YIPs, have proved more successful with these last organisms.

7.2 CLONING VECTORS FOR HIGHER PLANTS

There are important potential benefits of gene cloning using higher plants as the host organisms. Already gene cloning has resulted in experimental plants resistant to viruses and insects, and soon we can expect to see tomatoes with improved storage properties, and ornamental flowers in new colours and shades, thanks to the ability to introduce new genes into plants by gene cloning. There is real optimism that novel varieties of crops, with improved nutritional qualities and the ability to grow under adverse conditions, will be developed thanks to gene cloning.

Three types of vector system have been used with higher plants.

1. The Ti plasmid of *Agrobacterium tumefaciens*.
2. Plant viruses such as the caulimoviruses and geminiviruses.
3. Direct gene transfer using DNA fragments not attached to a plant cloning vector.

7.2.1 *AGROBACTERIUM TUMEFACIENS* – NATURE'S SMALLEST GENETIC ENGINEER

Although no naturally occurring plasmids are known in higher plants, one bacterial plasmid, the Ti plasmid of *A. tumefaciens*, is of great importance.

A. tumefaciens is a soil microorganism that causes crown gall disease in many species of dicotyledonous plants. Crown gall occurs when a wound on the stem allows *A. tumefaciens* bacteria to invade the plant. After infection the bacteria cause a cancerous proliferation of the stem tissue in the region of the crown (Figure 7.9).

The ability to cause crown gall disease is associated with the presence of the Ti (tumour Inducing) plasmid within the bacterial cell. This is a large (greater than 200 kb) plasmid that carries numerous genes involved in the infective process (Figure 7.10a). A remarkable feature of the Ti plasmid is that, after infection, part of the molecule is integrated into the plant chromosomal DNA (Figure 7.10b). This segment, called the **T-DNA**, is between 15 and 30 kb in size, depending on the strain. It is maintained in a

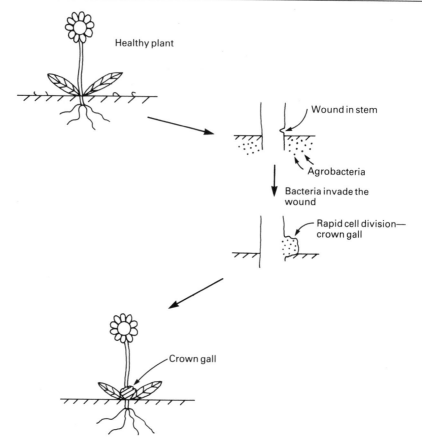

Healthy plant

Wound in stem

Agrobacteria

Bacteria invade the wound

Rapid cell division—crown gall

Crown gall

Figure 7.9 Crown gall disease.

stable form in the plant cell and is passed on to daughter cells as an integral part of the chromosomes. But the most remarkable feature of the Ti plasmid is that the T-DNA contains eight or so genes that are expressed in the plant cell and are responsible for the cancerous properties of the transformed cells. These genes also direct synthesis of unusual compounds, called opines, that the bacteria use as nutrients (Figure 7.10c). In short, *A. tumefaciens* genetically engineers the plant cell for its own purposes.

(a) Using the Ti plasmid to introduce new genes into a plant cell

It was realized very quickly that the Ti plasmid could be used to transport new genes into plant cells. All that would be necessary would be to insert the new genes into the T-DNA and then the bacterium could do the hard work of integrating them into the plant chromosomal DNA. In practice this has

(a) A Ti plasmid

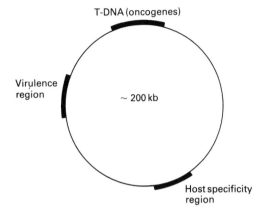

(b) Integration of the T-DNA into the plant genome

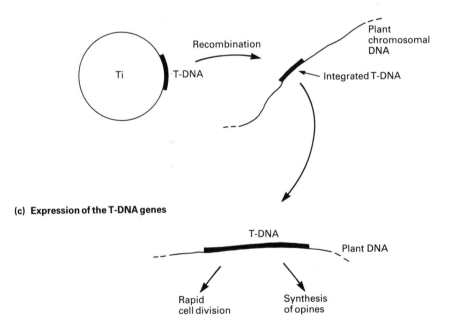

(c) Expression of the T-DNA genes

Figure 7.10 The Ti plasmid and its integration into the plant chromosomal DNA after *Agrobacterium tumefaciens* infection.

proved quite a tricky proposition, mainly because the large size of the Ti plasmid makes manipulation of the molecule very difficult.

The main problem is of course that a unique restriction site is an impossibility with a plasmid 200 kb in size. Novel strategies have to be developed for inserting new DNA into the plasmid. Two are in general use.

1. **The binary vector strategy** (Figure 7.11) is based on the observation that the T-DNA does not need to be physically attached to the rest of the Ti plasmid. A two-plasmid system, with the T-DNA on a relatively small molecule, and the rest of the plasmid in normal form, is just as effective at transforming plant cells. In fact some strains of *A. tumefaciens*, and related Agrobacteria, have natural binary plasmid systems. The T-DNA plasmid is small enough to have a unique restriction site and to be manipulated using standard techniques.
2. **The cointegration strategy** (Figure 7.12) uses an entirely new plasmid, based on pBR322 or a similar *E. coli* vector, but carrying a small portion of the T-DNA. The homology between the new molecule and the Ti plasmid means that if both are present in the same *A. tumefaciens* cell, then recombination can integrate the pBR plasmid into the T-DNA region. The gene to be cloned is therefore inserted into a unique restriction site on the small pBR plasmid, introduced into *A. tumefaciens* cells carrying a Ti plasmid, and the natural recombination process left to integrate the new gene into the T-DNA. Infection of the plant leads to insertion of the new gene, along with the rest of the T-DNA, into the plant chromosomes.

(b) Production of transformed plants with the Ti plasmid

If *A. tumefaciens* bacteria that contain an engineered Ti plasmid are introduced into a plant in the natural way, by infection of a wound in the stem, then only the cells in the resulting crown gall will possess the cloned gene

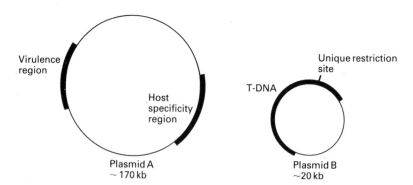

Plasmid A
~ 170 kb

Plasmid B
~20 kb

Figure 7.11 The binary vector strategy. Plasmids A and B complement each other when present together in the same *Agrobacterium tumefaciens* cell. The T-DNA carried by plasmid B is transferred to the plant chromosomal DNA by proteins coded by genes carried by plasmid A.

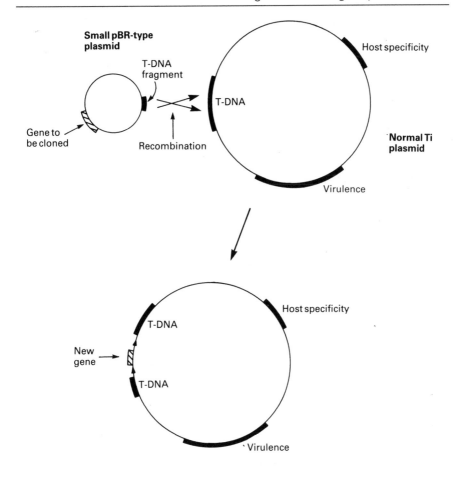

Figure 7.12 The cointegration strategy.

(Figure 7.13a). This is obviously of little value to the biotechnologist. Instead a way of introducing the new gene into every cell in the plant is needed.

The answer is to infect not the mature plant but a plant segment or a culture of plant cells in liquid medium (Figure 7.13b). These plant cells can be treated in the same way as microorganisms; for example they can be plated on to a selective medium in order to isolate transformants. A mature plant regenerated from transformed cells will contain the cloned gene in every cell and will pass the cloned gene to its offspring.

However, regeneration is not always possible. Firstly, it will occur only if the Ti vector has been **'disarmed'** by deletion of at least some of the genes that are normally responsible for the cancerous properties of transformed cells. A number of disarmed Ti cloning vectors are now available, examples

(a) Wound infection by recombinant *A.tumefaciens*

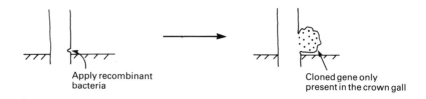

Apply recombinant
bacteria

Cloned gene only
present in the crown gall

(b) Transformation of cultured plant cells

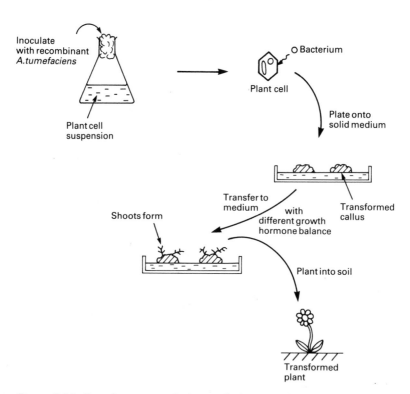

Inoculate
with recombinant
A.tumefaciens

Plant cell
suspension

Bacterium

Plant cell

Plate onto
solid medium

Transfer to
medium

with
different growth
hormone balance

Transformed
callus

Shoots form

Plant into soil

Transformed
plant

Figure 7.13 Transformation of plant cells by recombinant *Agrobacterium tumefaciens*. (a) Infection of a wound: transformed plant cells are present only in the crown gall. (b) Transformation of a cell suspension: all the cells in the resulting plant are transformed.

being Bin19, a binary vector (Figure 7.14a), and the cointegrate vector pGV3850 (Figure 7.14b). Despite lacking all the oncogenes present in the natural Ti plasmid, the T-DNA regions of these vectors are still transferred to the plant chromosomal DNA.

(a) Bin19

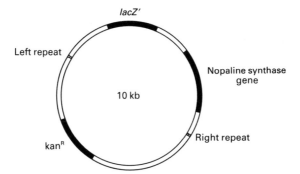

(b) pGV3850

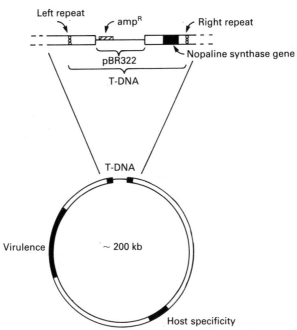

Figure 7.14 Two disarmed Ti cloning vectors. (a) Bin19, a binary vector, which has a number of cloning sites within the *lacZ'* gene. (b) pGV3850, a cointegrate vector which is used to shuttle genes originally inserted into pBR322 into a host plant. Neither plasmid contains any oncogenes but both have the left and right terminal repeats of the T-DNA region.

A second problem with regeneration of mature plants from transformed cells is that the efficiency of the process depends very much on the particular species involved. Many plants can be regenerated with ease but others are much more difficult. Unfortunately amongst the difficult ones are many of the most important crop plants, including monocotyledonous species such as cereals (barley, wheat and maize, for example) and grasses (rice and others). However, regeneration is a technical problem that will probably be solved by experimentation. A potentially more serious difficulty is presented by the fact that in nature *A. tumefaciens* infects only dicotyledonous plants: monocots are outside of the normal host range. For several years it was thought that there was an effective barrier to infection of monocots with *A. tumefaciens* and it seemed that the Ti plasmid could not be used to clone genes into cereals and grasses. Now these fears appear unduly pessimistic as transfer of T-DNA to chromosomes of a number of monocots has been achieved in the laboratory. As yet the process is less efficient than with the natural host plants but further research into the infection process should solve the remaining problems.

7.2.2 GENE CLONING USING PLANT VIRUSES

Although most plants are subject to viral infection, the vast majority of these viruses have genomes not of DNA but of RNA. These are not so useful as potential cloning vectors because manipulations with RNA are rather more difficult to carry out.

Only two classes of DNA viruses are known to infect higher plants. The first of these are the **caulimoviruses**, of which cauliflower mosaic virus (CaMV) is a typical example.

CaMV has attracted considerable interest as the possible basis for a cloning vector. The genomes of the caulimoviruses are all quite small (8 kb or so), and that of CaMV has been completely characterized by DNA sequencing, revealing at least six closely packed genes and a single intergenic region (Figure 7.15). Additional DNA can be inserted into this intergenic region without destroying the infective capability of the virus. Caulimoviruses also have the useful property of being able to spread through a plant from a single infection point. Inoculation of a small number of virus particles on to the surface of a leaf will result eventually in all the cells in the plant becoming infected. Transformed plants can therefore be obtained without the need for regeneration from cell cultures.

Although a CaMV-based vector has successfully been used to clone a new gene into turnip plants, two major problems have limited its general usefulness. The first is that the total size of the CaMV genome is, like λ, constrained by the need to package it into its protein coat. In the case of CaMV only 200 to 300 additional nucleotides can be accommodated. Even if sections of the CaMV genome were to be deleted, the capacity for carrying inserted DNA would still be very limited.

The second problem with CaMV is its host range, which is much more

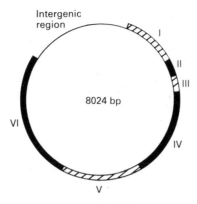

Figure 7.15 The cauliflower mosaic virus (CaMV) genome.

restricted than that of the Ti plasmid. Caulimoviruses naturally infect only a small number of species, primarily the brassicas such as cauliflower and turnip. At the moment there seems to be little prospect of extending this host range to include cereals and grasses.

The second group of plant DNA viruses, the **geminiviruses**, also have a limited host range, but with these viruses the hosts include important crop plants such as maize and wheat. Geminiviruses have small genomes, 5 to 6 kb in size, typically comprising not one but two DNA molecules. Infection of a plant can be achieved by surface inoculation, at least with most of the viral types. These useful features are counterbalanced by two disadvantages that have so far prevented the viruses from being developed as cloning vectors. The first problem is that during the natural infection cycle geminiviral genomes undergo rearrangements and deletions, which would scramble up any additional DNA that had been inserted. The second problem is that many of these viruses cause damaging infections to crops, so the use of a geminivirus cloning vector would be subject to stringent controls, to avoid the danger of the vector escaping from the host and infecting natural plant populations. The geminiviruses are, however, currently looked on as a better proposition than the caulimoviruses, and future research may well produce a stable, disarmed gemini-vector.

7.2.3 DIRECT GENE TRANSFER

The Ti plasmid and the two types of plant DNA virus are natural systems that molecular biologists have modified, or are trying to modify, to allow new genes to be introduced into plants. Each type of system is hindered by natural constraints that could limit its usefulness. Could an artificial system be of greater value?

The answer is yes, possibly. A bacterial plasmid such as pBR322 can be introduced into plant protoplasts (p. 100), and although it will not replicate on its own it may be integrated by recombination into one of the plant chromosomes. This event is distinct from the recombination responsible for chromosomal integration of a yeast vector (p. 130), as there is no region of homology between pBR322 and the plant DNA. The process is not well understood, but appears to occur randomly, resulting in integration at any position in any of the plant chromosomes (Figure 7.16). One requirement is that the plasmid must be supercoiled.

If the plasmid introduced into the plant protoplasts carries a gene, then that gene will be integrated into the plant chromosomal DNA, where it may be maintained, being passed to the progeny just as though it were a normal plant gene. This cloning system is called **direct gene transfer**.

Direct gene transfer has been successful in introducing new genes into protoplasts of a variety of plants, both dicotyledonous and monocotyledonous. The one major problem with the system is that regeneration of a mature plant is even more difficult from protoplasts than it is from cultured cells; although new cell walls form spontaneously, redifferentiation to produce plants often will not occur. However, there is no apparent host specificity for direct gene transfer and the technique seems applicable to all plant species. This last factor may make direct gene transfer the system of choice for the genetic engineering of crop plants.

7.3 CLONING VECTORS FOR MAMMALIAN CELLS

Considerable effort has been put into the development of vector systems that will allow genes to be cloned in mammalian cells. Much of this work has been carried out by biotechnology companies who subsequently use the cloning system for a specific applied purpose, several examples of which will be described in Chapter 12. Unfortunately, because of the profit angle, precise details of the cloning techniques are often not released to the general scientific community, an undesirable state of affairs that lends mystery and confusion to the field.

The simplest mammalian cloning vectors are based on bacterial plasmids such as pBR322, allowing manipulations to be carried out in E. coli, and rely on integration into chromosomal DNA for their stability after introduction into the mammalian host cell. The concept is similar to direct gene transfer in plants and will not be described in detail. In addition there is a range of vectors derived from mammalian viruses, two of which will be considered here.

7.3.1 VECTORS BASED ON SIMIAN VIRUS 40

Of all the mammalian viruses used for gene cloning, simian virus 40 (SV40) has received most attention. The virus is capable of infecting several mammalian species, following a lytic cycle in some hosts and a lysogenic cycle in others. The genome is 5.2 kb in size (Figure 7.17a) and contains two

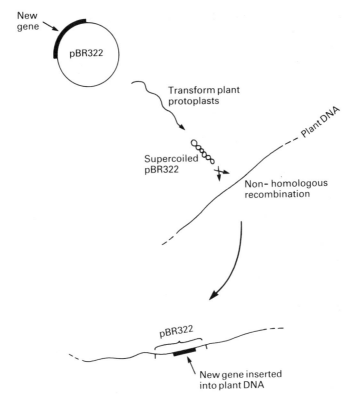

New
gene

pBR322

Transform plant
protoplasts

Plant DNA

Supercoiled
pBR322

Non- homologous
recombination

pBR322

New gene inserted
into plant DNA

Figure 7.16 Direct gene transfer.

sets of genes, the 'early' genes, expressed early in the infection cycle and
coding for proteins involved in viral DNA replication, and the 'late' genes,
coding for viral capsid proteins.

SV40 suffers from the same problem as λ, in that packaging constraints
limit the amount of new DNA that can be inserted into the genome. As the
SV40 molecule is quite small only a few hundred extra nucleotides can be
tolerated. The most successful SV40-based vectors are therefore deleted
versions of the normal genome, in which SV40 genes are replaced by the
new DNA. An example is provided by SVGT-5 (Figure 7.17b), which was the
first SV40 vector to be reported and was used to clone the rabbit β-globin
gene.

7.3.2 VECTORS BASED ON PAPILLOMAVIRUSES

With SV40 vectors the host cell is usually killed after a short time by the lytic
action of the virus. To obtain stable transformed mammalian cells a different
type of virus must be used.

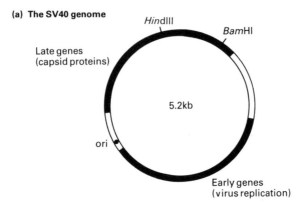

(a) The SV40 genome

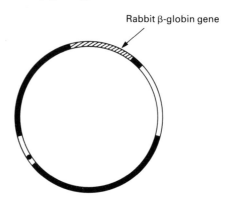

(b) SVGT-5 with the inserted rabbit β-globin gene

Figure 7.17 SV40 and an example of its use as a cloning vector. To clone the rabbit β-globin gene the *Hind*III to *Bam*HI restriction fragment was deleted (resulting in SVGT-5) and replaced with the rabbit gene.

Papillomaviruses such as **bovine papillomavirus** (BPV) have been the most successful vectors for obtaining permanent transformants. BPV primarily causes warts on cattle, but can also infect other species. In mouse cells BPV takes the form of a multicopy plasmid, with about 100 molecules present per cell. It does not cause death of the host cell, and BPV molecules are passed to daughter cells on cell division. Shuttle vectors consisting of BPV and pBR322 sequences, and capable of replication in both mouse and bacterial cells, have been constructed and show great potential for the introduction of large pieces of DNA into mammalian cells.

FURTHER READING

Parent, S. A. *et al.* (1985) Vector systems for the expression, analysis and cloning of DNA sequences in *S. cerevisiae*. *Yeast*, **1**, 83–138 – details of different yeast cloning vectors.

Broach, J. R. (1982) The yeast 2 μm circle. *Cell*, **28**, 203–4.

Burke, D. T. *et al.* (1987) Cloning of large segments of exogenous DNA into yeast by means of artificial chromosome vectors. *Science*, **236**, 806–12.

Chilton, M. D. (1983) A vector for introducing new genes into plants. *Scientific American*, **248** (June), 50–9 – the Ti plasmid.

Bevan, M. (1984) Binary *Agrobacterium* vectors for plant transformation. *Nucleic Acids Research*, **12**, 8711–21.

Brisson, N. *et al.* (1984) Expression of a bacterial gene in plants by using a viral vector. *Nature*, **310**, 511–14 – a cloning experiment with CaMV.

Davies, J. W. and Stanley, J. (1989) Geminivirus genes and vectors. *Trends in Genetics*, **5**, 77–81.

Paszkowski, J. *et al.* (1984) Direct gene transfer to plants. *EMBO Journal*, **3**, 2717–22.

Hamer, D. H. and Leder, P. (1979) Expression of the chromosomal mouse β-maj-globin gene cloned in SV40. *Nature*, **281**, 35–40.

Part Two

The Applications of Cloning in Gene Analysis

How to obtain a clone of a specific gene

8

So far in this book gene cloning has been considered somewhat as an end in itself, with no attempt made to put the technique into context with molecular biological research as a whole. In particular, the examples of gene cloning used to illustrate the methodology have made use of 'DNA fragments' of unspecified origin and indeterminate importance. The astute reader will by now be wondering how construction of recombinant DNA molecules and transformation of living cells are used in the real world to obtain clones of specific genes and to provide information of importance to molecular biology and biotechnology.

In this part of the book, the role and relevance of gene cloning should become clear. Firstly, in this chapter, the methods available for obtaining a clone of an individual, specified gene will be described. This is in fact the critical test of a gene cloning experiment – success or failure often depends on whether or not a strategy can be devised by which clones of the desired gene can be selected directly, or alternatively, distinguished from other recombinants. Once this problem has been resolved, and a clone has been obtained, the molecular biologist is able to make use of a wide variety of different techniques which will extract information about the gene. The most important of these will be described in Chapters 9 and 10.

8.1 THE PROBLEM OF SELECTION

The problem faced by the molecular biologist wishing to obtain a clone of a single, specified gene was illustrated in Figure 1.3. Even the simplest organisms, such as *E. coli*, contain several thousand genes, and a restriction digest of total cell DNA will produce not only the fragment carrying the desired gene, but also many other fragments carrying all the other genes (Figure 8.1a). During the ligation reaction there will of course be no selection for an individual fragment: numerous different recombinant DNA molecules will be produced, all containing different pieces of DNA (Figure 8.1b). Consequently a variety of recombinant clones will be obtained after

(a) Restriction of a large DNA molecule

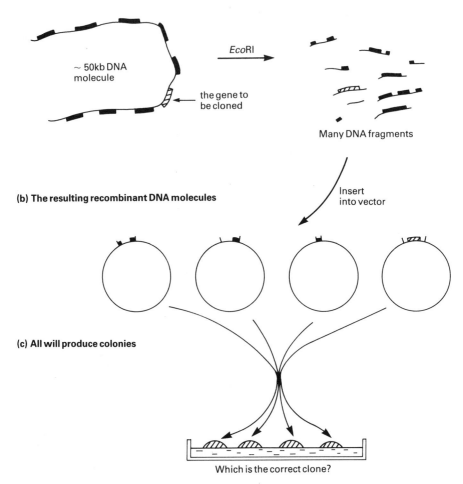

(b) The resulting recombinant DNA molecules

(c) All will produce colonies

Which is the correct clone?

Figure 8.1 The problem of selection.

transformation and plating out (Figure 8.1c). Somehow the correct one must be identified.

8.1.1 THERE ARE TWO BASIC STRATEGIES FOR OBTAINING THE CLONE YOU WANT

Although there are many different procedures by which the desired clone can be obtained, all are variations on two basic themes.

1. **Direct selection for the desired gene** (Figure 8.2a), which means that the

(a) Direct selection

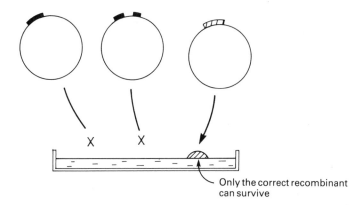

X   X

Only the correct recombinant
can survive

(b) Clone identification

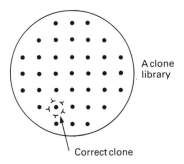

A clone
library

Correct clone

Figure 8.2 The basic strategies that can be used to obtain a particular clone.
(a) Direct selection. (b) Identification of the desired recombinant from a clone library.

cloning experiment is designed in such a way that the only clones that are obtained are clones of the required gene. Almost invariably, selection occurs at the plating-out stage.
2. **Identification of the clone from a gene library** (Figure 8.2b), which entails an initial **'shotgun'** cloning experiment, to produce a clone library representing all or most of the genes present in the cell, followed by analysis of the individual clones to identify the correct one.

In general terms, direct selection is the preferred method, as it is quick and usually unambiguous. However, as we shall see, it is not applicable to all genes. Techniques for clone identification are therefore very important, especially as complete genomic libraries of many organisms are now available.

8.2 DIRECT SELECTION

To be able to select for a cloned gene it is necessary to plate the transformants on to an agar medium on which only the desired recombinants, and no others, can grow. The only colonies that will appear will therefore comprise cells that contain the desired recombinant DNA molecule.

The simplest example of direct selection occurs when the desired gene specifies resistance to an antibiotic. As an example we will consider an experiment to clone the gene for kanamycin resistance from plasmid R6.5. This plasmid in fact carries genes for resistances to four antibiotics: kanamycin, chloramphenicol, streptomycin and sulphonamide. The kanamycin resistance gene lies within one of the 13 *Eco*RI fragments (Figure 8.3a).

To clone this gene the *Eco*RI fragments of R6.5 would be inserted into the *Eco*RI site of a vector such as pBR322. The ligated mix will comprise many copies of 13 different recombinant DNA molecules, one set of which will carry the gene for kanamycin resistance (Figure 8.3b).

Insertional inactivation cannot be used to select recombinants when the *Eco*RI site of pBR322 is used. This is because this site does not lie in either the ampicillin or the tetracycline resistance genes of this plasmid (Figure 6.1). But this is immaterial for cloning the kanamycin resistance gene as in this case the cloned gene can be used as the selectable marker. Transformants are plated on to kanamycin agar, on which the only cells able to survive and produce colonies will be those recombinants that contain the cloned kanamycin resistance gene (Figure 8.3c).

8.2.1 MARKER RESCUE EXTENDS THE SCOPE OF DIRECT SELECTION

Direct selection would be very limited indeed if it could be used only for cloning antibiotic resistance genes. Fortunately the technique can be extended by making use of mutant strains of *E. coli* as the hosts for transformation.

As an example, consider an experiment to clone the gene *trpA* from *E. coli*. This gene codes for the enzyme tryptophan synthase, which is involved in biosynthesis of the essential amino acid tryptophan. A mutant strain of *E. coli* that has a non-functional *trpA* gene is called *trpA⁻*, and is able to survive only if tryptophan is added to the growth medium. *E. coli trpA⁻* is therefore another example of an auxotroph (p. 129).

This *E. coli* mutant can be used to clone the correct version of the *trpA* gene. Total DNA is first purified from a normal (wild-type) strain of the bacterium. Digestion with a restriction endonuclease, followed by ligation into a vector, will produce numerous recombinant DNA molecules, one of which may, with luck, carry an intact copy of the *trpA* gene (Figure 8.4a). This will of course be the functional gene as it has been obtained from the wild-type strain.

The ligation mixture is now used to transform the auxotrophic *E. coli trpA⁻*

(a) Plasmid R6.5

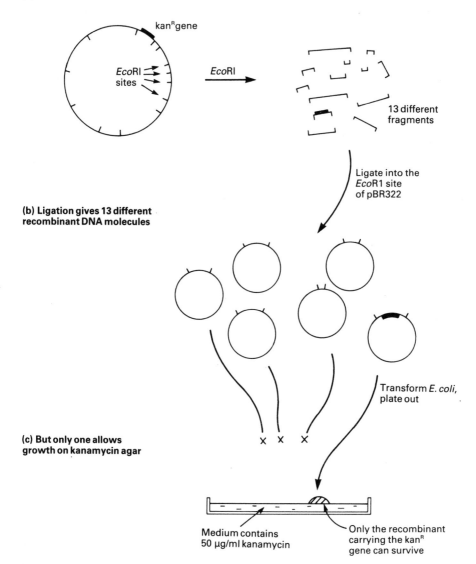

(b) Ligation gives 13 different recombinant DNA molecules

(c) But only one allows growth on kanamycin agar

Figure 8.3 Direct selection for the cloned R6.5 kanamycin resistance (kan^R) gene.

cells (Figure 8.4b). The vast majority of the resulting transformants will be auxotrophic, but a few will now have the plasmid-borne copy of the correct *trpA* gene. These recombinants will be non-auxotrophic – they will no longer require tryptophan as the cloned gene will be able to direct production of tryptophan synthase (Figure 8.4c). Direct selection is therefore performed by plating transformants on to minimal medium, which lacks any added

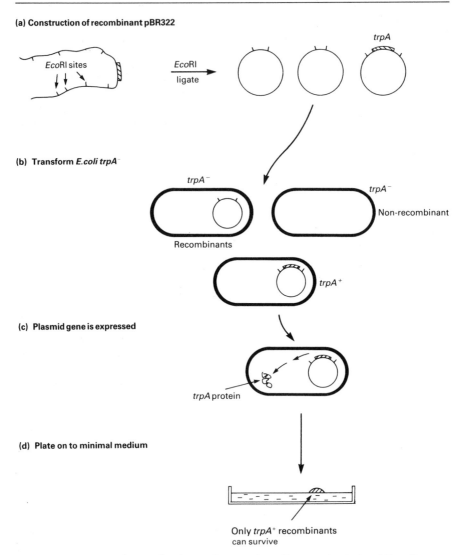

(a) Construction of recombinant pBR322

EcoRI sites

$\xrightarrow[\text{ligate}]{EcoRI}$

trpA

(b) Transform E.coli trpA⁻

trpA⁻

Recombinants

trpA⁻

Non-recombinant

trpA⁺

(c) Plasmid gene is expressed

trpA protein

(d) Plate on to minimal medium

Only trpA⁺ recombinants
can survive

Figure 8.4 Direct selection for the *trpA* gene cloned in a *trpA⁻* strain of *E. coli*.

supplements, and in particular has no tryptophan (Figure 8.4d). Auxotrophs will not grow on minimal medium, so the only colonies that will
appear will be recombinants that contain the cloned *trpA* gene.

8.2.2 THE SCOPE AND LIMITATIONS OF MARKER RESCUE

Although marker rescue can be used to obtain clones of many genes, the
technique is subject to two limitations.

1. A mutant strain must be available for the gene in question.
2. A medium on which only the wild-type can survive is needed.

In general terms, marker rescue is applicable for genes that code for biosynthetic enzymes, as these can be selected on minimal medium in the manner described for *trpA*. However, the technique is not limited to *E. coli* nor even bacteria; auxotrophic strains of yeast and filamentous fungi are also available, and marker rescue has been used to select genes cloned into these organisms.

In addition, *E. coli* auxotrophs can be used as hosts for the selection of some genes from other organisms. Often there is sufficient similarity between equivalent enzymes from different bacteria, or even from yeast, for the foreign enzyme to function in *E. coli*, so that the cloned gene will be able to transform the host to wild-type.

8.3 IDENTIFICATION OF A CLONE FROM A GENE LIBRARY

Although marker rescue is a powerful technique it is not all-embracing and there are many important genes that cannot be selected by this method. Many bacterial mutants are not auxotrophs, so the mutant and wild-type strains cannot be distinguished by plating on to minimal or any other special medium. In addition, neither marker rescue nor any other direct selection method will be of much use in providing bacterial clones of genes from higher organisms (i.e. animals and plants), as in these cases the differences are usually so great that the foreign enzymes will not function in the bacterial cell.

The alternative strategy must therefore be considered. This is where a large number of different clones are obtained and the desired one identified in some way.

8.3.1 GENE LIBRARIES

Before looking at the methods used to identify individual clones, the library itself must be considered. A genomic library (p. 126) is a collection of clones sufficient in number to be likely to contain every single gene present in a particular organism. Genomic libraries are prepared by purifying total cell DNA, and then making a partial restriction digest, resulting in fragments that can be cloned into a suitable vector (Figure 8.5), usually a λ replacement vector or a cosmid.

For bacteria, yeast and fungi, the number of clones needed for a complete genomic library is not so large as to be unmanageable (Table 6.1). For plants and animals, though, a complete library will contain so many different clones that identification of the desired one may prove a mammoth task. With these organisms a second type of library, specific not to the whole organism but to a particular cell type, may be more useful.

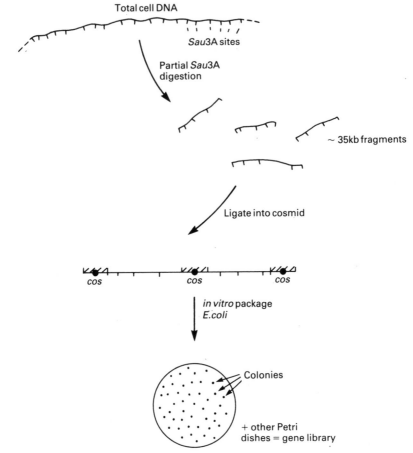

Figure 8.5 Preparation of a gene library in a cosmid vector.

8.3.2 NOT ALL GENES ARE EXPRESSED AT THE SAME TIME

A characteristic of most multicellular organisms is specialization of individual cells. A human being, for example, is made up of a large number of different cell types – brain cells, blood cells, liver cells, etc. Each cell will of course contain the same complement of genes, but in different cell types different sets of genes will be switched on, others will be silent (Figure 8.6).

The fact that only relatively few genes are expressed in any one type of cell can be utilized in preparation of a library if the material that is cloned is not DNA but **messenger RNA (mRNA)**. Only those genes that are being expressed are transcribed into mRNA, and so if mRNA is used as the starting material then the resulting clones will comprise only a selection of the total number of genes in the cell.

Cell type A

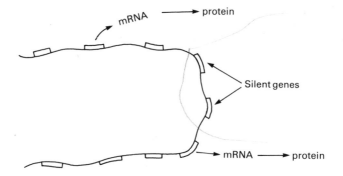

Cell type B

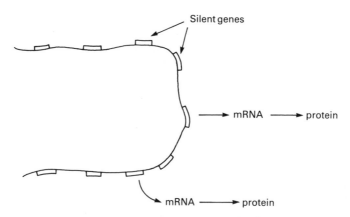

Figure 8.6 Different genes are expressed in different types of cell.

A cloning method that uses mRNA would be particularly useful if the desired gene is expressed at a high rate in an individual cell type. For example, the gene for gliadin, one of the nutritionally important proteins present in wheat, is expressed at a very high level in the cells of developing wheat seeds. In these cells over 30% of the total mRNA specifies gliadin. Clearly if we could clone the mRNA from wheat seeds then we would obtain a large number of clones specific for gliadin.

8.3.3 mRNA CAN BE CLONED AS COMPLEMENTARY DNA

mRNA cannot itself be ligated into a cloning vector. However, mRNA can be converted into DNA by **complementary DNA (cDNA)** synthesis.

The key to this method is the enzyme reverse transcriptase (p. 56) which will synthesize a DNA polynucleotide strand complementary to an existing RNA strand (Figure 8.7a). Once the cDNA strand has been synthesized the RNA member of the hybrid molecule can be partially degraded by treating with ribonuclease H (Figure 8.7b). The remaining RNA fragments then serve as primers (p. 54) for DNA polymerase I, which will synthesize the second cDNA strand (Figure 8.7c), resulting in a double-stranded DNA fragment that can be ligated into a vector and cloned (Figure 8.7d).

The resulting cDNA clones will be representative of the mRNA present in the original preparation. In the case of mRNA prepared from wheat seeds, the cDNA library will contain a large proportion of clones representing gliadin mRNA (Figure 8.7e). Other clones will also be present, but locating the cloned gliadin cDNA will be a much easier process than identifying the equivalent gene from a complete wheat genomic library.

8.4 METHODS FOR CLONE IDENTIFICATION

Once a suitable library has been prepared, a number of procedures can be employed to attempt identification of the desired clone. Although a few of these procedures are based on detection of the translation product of the cloned gene, it is usually easier to identify directly the correct recombinant DNA molecule. This can be achieved by the important technique of **hybridization probing**.

8.4.1 COMPLEMENTARY NUCLEIC ACID STRANDS WILL HYBRIDIZE TO EACH OTHER

Any two single-stranded nucleic acid molecules have the potential to form base pairs with one another. With most pairs of molecules the resulting hybrid structures will be unstable, as only a small number of individual interstrand bonds will be formed (Figure 8.8a). However, if the polynucleotides are complementary (p. 22) then extensive base pairing will occur to form a stable double-stranded molecule (Figure 8.8b). Not only can this occur between single-stranded DNA molecules to form the DNA double helix, but also between single-stranded RNA molecules and between combinations of one DNA and one RNA strand (Figure 8.8c).

Nucleic acid hybridization can be used to identify a particular recombinant clone if a DNA or RNA probe, complementary to the desired gene, is available. The exact nature of the probe will be discussed later in the chapter. First we must consider the technique itself.

8.4.2 COLONY AND PLAQUE HYBRIDIZATION PROBING

Hybridization probing can be used to identify recombinant DNA molecules contained in either bacterial colonies or bacteriophage plaques. Thanks to

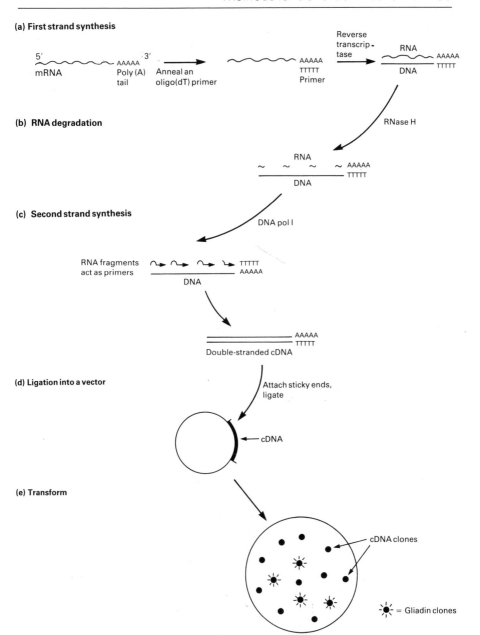

(a) First strand synthesis

(b) RNA degradation

(c) Second strand synthesis

(d) Ligation into a vector

(e) Transform

Figure 8.7 One possible scheme for cDNA cloning. See the text for details.

(a) An unstable hybrid

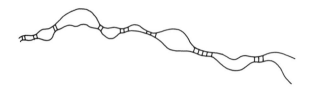

(b) A stable hybrid

Short non-complementary regions
do not affect overall stability

(c) A DNA–RNA hybrid

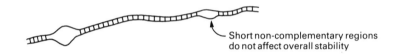

DNA
RNA

Figure 8.8 Nucleic acid hybridization. (a) An unstable hybrid molecule formed between two non-homologous DNA strands. (b) A stable hybrid formed between two complementary strands. (c) A DNA–RNA hybrid, such as may be formed between a gene and its transcript.

innovative techniques developed in the late 1970s it is no longer necessary to purify each recombinant molecule. Instead an *in situ* probing method is used.

First the colonies or plaques are transferred to a nitrocellulose or nylon membrane (Figure 8.9a) and then treated to remove all contaminating material, leaving just DNA (Figure 8.9b). Usually this treatment also results in denaturation of the DNA molecules, so that the hydrogen bonds between individual strands in the double helix are broken. These single-stranded molecules can then be bound tightly to the membrane by a short period at 80°C if a nitrocellulose membrane is being used, or with a nylon membrane by ultraviolet irradiation. The molecules will in fact be attached to the membrane through their sugar–phosphate backbones, so the bases will be free to pair with complementary nucleic acid molecules.

The probe must now be labelled, denatured by heating, and applied to the membrane in a solution of chemicals that promote nucleic acid hybridization

(a) Transfer colonies to nitrocellulose or nylon

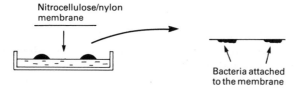

(b) Degrade cells, purify DNA

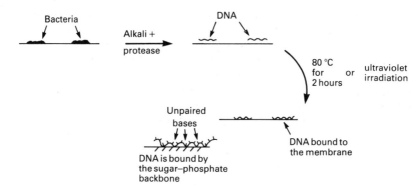

(c) Probe with labelled DNA

(d) The resulting autoradiograph

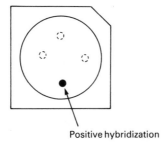

Positive hybridization

Figure 8.9 Colony hybridization probing with a radioactively labelled probe.

(Figure 8.9c). After a period to allow hybridization to take place, the filter is washed to remove unbound probe, dried, and the positions of the bound probe detected (Figure 8.9d).

Traditionally the probe is labelled with a radioactive nucleotide, either by nick translation or end-filling (p. 68), or alternatively by **random priming** (Figure 8.10), a technique that results in a probe with higher activity and therefore able to detect smaller amounts of bound complementary DNA. With these methods the position of the hybridization signal is determined by autoradiography. Radioactive labelling methods are starting to fall out of favour, however, partly because of the hazard to the researcher and partly because of the problems associated with disposal of radioactive waste. The hybridization probe may therefore be labelled in a non-radioactive manner. A number of methods have been developed, two of which are illustrated in Figure 8.11. The first makes use of dUTP nucleotides modified by reaction with **biotin**, an organic molecule that has a high affinity for a protein called **avidin**. After hybridization the positions of the bound biotinylated probe can be determined by washing with avidin coupled to a fluorescent marker (Figure 8.11a). This method is as sensitive as radioactive probing and is

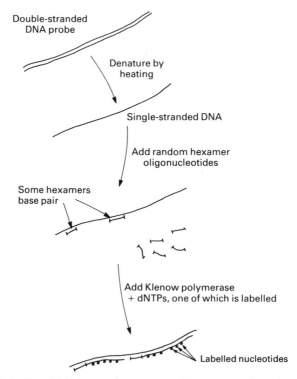

Figure 8.10 Labelling DNA by random priming. The mixture of random hexamers will be sufficiently complex to include at least a few molecules that will base pair to the probe.

(a) Labelling with a biotinylated nucleotide

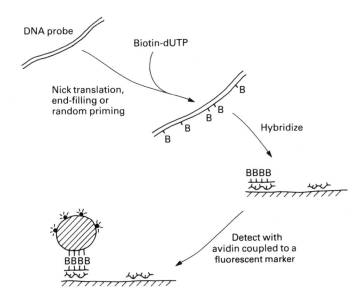

(b) Labelling with horseradish peroxidase

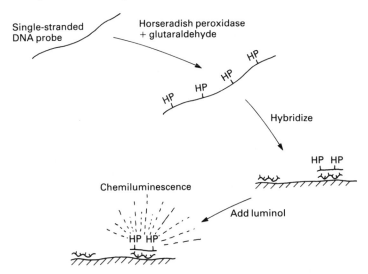

Figure 8.11 Two methods for the non-radioactive labelling of DNA probes.

becoming increasingly popular. The same goes for the second procedure for non-radioactive hybridization probing, where the probe DNA is complexed with the enzyme **horseradish peroxidase**, and is detected through the enzyme's ability to degrade luminol with the emission of chemiluminescence (Figure 8.11b). The signal can be recorded on normal photographic film in a manner analogous to autoradiography.

8.4.3 EXAMPLES OF THE PRACTICAL USE OF HYBRIDIZATION PROBING

Clearly the success of colony or plaque hybridization as a means of identifying a particular recombinant clone hangs on the availability of a DNA molecule that can be used as a probe. This probe must share at least a part of the sequence of the cloned gene. If the gene itself is not available (which presumably will be the case if the aim of the experiment is to provide a clone of it) then what can be used as the probe?

In practice the nature of the probe is determined by the information available about the desired gene. We will consider three possibilities.

1. Where the desired gene is expressed at a high level in a cell type from which a cDNA clone library has been prepared.
2. Where the amino acid sequence of the protein coded by the gene is completely or partially known.
3. Where the gene is a member of a family of related genes.

(a) Abundancy probing to analyse a cDNA library

As described earlier in this chapter, a cDNA library is often prepared in order to obtain a clone of a gene expressed at a relatively high level in a particular cell type. In the example of a cDNA library from developing wheat seeds, a large proportion of the clones will be copies of the mRNA transcripts of the gliadin gene (Figure 8.7e).

Identification of the gliadin clones is simply a case of using individual cDNAs from the library to probe all the other members of the library (Figure 8.12). A clone is selected at random and the recombinant DNA molecule purified, labelled and used to probe the remaining clones. This is repeated with different clones as probes until one that hybridizes to a large proportion of the library is obtained. This abundant cDNA will be considered a possible gliadin clone and analysed in greater detail (e.g. by DNA sequencing and isolation of the translation product) to confirm the identification.

(b) Oligonucleotide probes for genes whose translation products have been characterized

Often the gene to be cloned will code for a protein that has already been studied in some detail. In particular the amino acid sequence of the protein

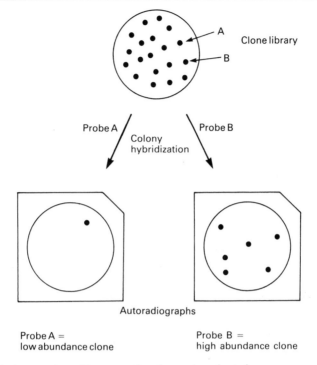

Figure 8.12 Probing within a library to identify an abundant clone.

may have been determined, using sequencing techniques that have been available for over 30 years. If the amino acid sequence is known, then it is possible to use the genetic code to predict the nucleotide sequence of the relevant gene. This prediction will always be an approximation, as only methionine and tryptophan can be assigned unambiguously to triplet codons; all other amino acids are coded by at least two codons each. Nevertheless, in most cases, the different codons for an individual amino acid are related. Alanine, for example, is coded by GCA, GCT, GCG and GCC, so two out of the three nucleotides of the triplet coding for alanine can be predicted with certainty.

As an example to clarify how these predictions are made, consider cytochrome c, a protein that plays an important role in the respiratory chain of all aerobic organisms. The cytochrome c protein from yeast was sequenced in 1963, with the result shown in Figure 8.13. This sequence contains a segment, starting at amino acid 59, that runs Trp–Asp–Glu–Asn–Asn–Met. The genetic code states that this hexapeptide will be coded by TGG–GA$_C^T$–GA$_G^A$–AA$_C^T$–AA$_C^T$–ATG. Although this represents a total of 16 different possible sequences, 14 of the 18 nucleotides can in fact be predicted with certainty.

```
                                                              15
GLY – SER –  ALA – LYS – LYS – GLY –ALA – THR – LEU – PHE  – LYS – THR –ARG –CYS – GLU –
                                                              30
LEU – CYS –  HIS – THR – VAL – GLU – LYS – GLY – GLY – PRO – HIS  – LYS – VAL – GLY – PRO –
                                                              45
ASN – LEU –  HIS – GLY – ILE  – PHE – GLY – ARG – HIS – SER  – GLY – GLN –ALA –GLN – GLY –
                                                              60
TYR – SER –  TYR – THR – ASP – ALA – ASN – ILE  – LYS – LYS  – ASN – VAL – LEU – TRP – ASP –
                                                              75
GLU – ASN – ASN – MET – SER – GLU – TYR – LEU – THR – ASN – PRO – LYS – LYS – TYR – ILE –
                                                              90
PRO – GLY – THR – LYS  –MET – ALA – PHE – GLY – GLY – LEU –  LYS – LYS – GLU – LYS – ASP–
                                                              103
ARG – ASN – ASP – LEU –  ILE  – THR – TYR  – LEU – LYS – LYS  – ALA – CYS – GLU
```

Figure 8.13 The amino acid sequence of yeast cytochrome c. The hexapeptide that is italicized is the one used to illustrate how a nucleotide sequence can be predicted from an amino acid sequence.

For many years this type of prediction was of limited academic interest. But nowadays short oligonucleotides of predetermined sequence can be synthesized in the laboratory (Figure 8.14). An oligonucleotide probe can therefore be constructed according to the predicted nucleotide sequence, and this probe may be able to identify the gene coding for the protein in question. In the example of yeast cytochrome c, the 16 possible oligonucleotides that can code for Trp–Asp–Glu–Asn–Asn–Met would be synthesized, either separately or as a pool, and then used to probe a yeast genomic or cDNA library (Figure 8.15). The probe will share some nucleotide homology with the cytochrome c gene, and this may be sufficient to provide an unambiguous hybridization signal. Even if more than one clone gives positive hybridization, reprobing with an oligonucleotide predicted by a different segment of the cytochrome c amino acid sequence should provide a definite result (Figure 8.15b). However, the segment of the protein used for nucleotide sequence prediction must be chosen with care: the hexapeptide Ser–Glu–Tyr–Leu–Thr–Asn, which immediately follows our first choice, could be coded by several thousand different 18-nucleotide sequences, clearly an unsuitable choice for a synthetic probe.

(c) Heterologous probing allows related genes to be identified

Often a substantial amount of nucleotide **homology** is seen when two genes for the same protein, but from different organisms, are compared, a reflection of the conservation of gene structure during evolution. Frequently, two genes from related organisms will be sufficiently homologous for a single-stranded probe prepared from one gene to form a stable hybrid with the second gene. Although the two molecules will not be entirely complementary, enough base pairs will be formed to produce a stable structure (Figure 8.16a).

Heterologous probing makes use of hybridization between related sequences for clone identification. For example, the yeast cytochrome c

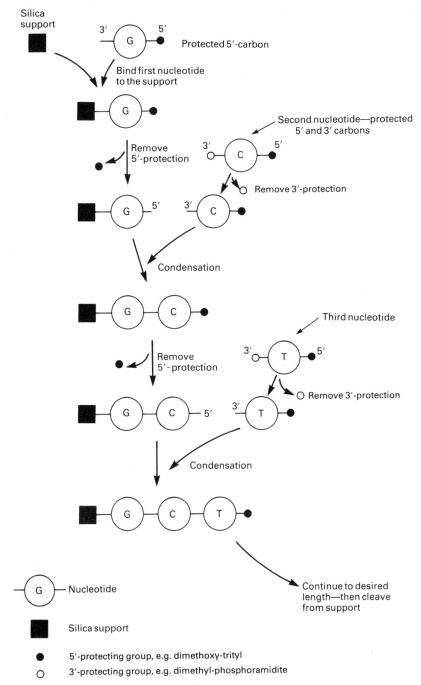

Figure 8.14 A simplified scheme for oligonucleotide synthesis. The protecting groups attached to the 3′ and 5′ termini prevent reactions between individual mononucleotides. By carefully controlling the time at which the protecting groups are removed, mononucleotides can be added one by one to the growing oligonucleotide.

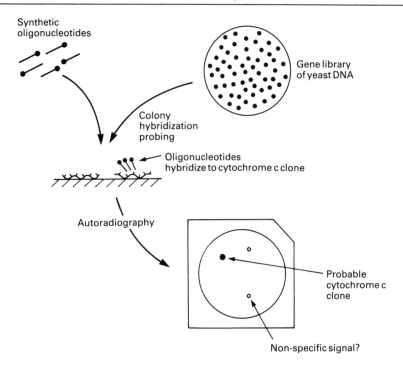

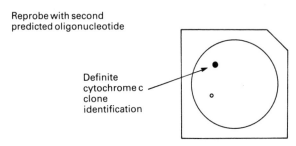

Figure 8.15 The use of a synthetic oligonucleotide to identify a clone of the yeast cytochrome c gene.

gene, identified in the previous section by oligonucleotide probing, could itself by used as a hybridization probe to identify cytochrome c genes in clone libraries of other organisms. A probe prepared from the yeast gene will not be entirely complementary to the gene from, say *Neurospora crassa*, but sufficient base pairing should occur for a hybrid to be formed and be detected by autoradiography (Figure 8.16b). The experimental conditions would in fact be modified so that the heterologous structure is not destabilized and lost before autoradiography.

(a) A hybrid between two related DNA strands

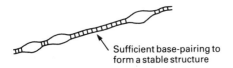

Sufficient base-pairing to
form a stable structure

(b) Heterologous probing between species

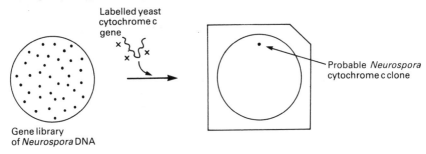

Labelled yeast
cytochrome c
gene

Gene library
of *Neurospora* DNA

Probable *Neurospora*
cytochrome c clone

(c) Heterologous probing within a species

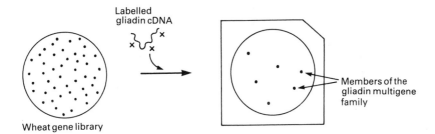

Labelled
gliadin cDNA

Wheat gene library

Members of the
gliadin multigene
family

Figure 8.16 Heterologous probing.

Heterologous probing can also identify related genes in the **same** organism. If the wheat gliadin cDNA clone, identified earlier in the chapter by abundancy probing, is used to probe a genomic library, then it will hybridize not only to its own gene, but to a variety of other genes as well (Figure 8.16c). These will all be related to the gliadin cDNA, but will have slightly different nucleotide sequences. This is because the wheat gliadins form a complex group of related proteins that are coded by the members of a **multigene family**. Once one gene in the family has been cloned, then all the other members can be isolated by heterologous probing.

8.4.4 IDENTIFICATION METHODS BASED ON DETECTION OF THE TRANSLATION PRODUCT OF THE CLONED GENE

Hybridization probing is usually the preferred method for identification of a particular recombinant from a clone library. The technique is easy to perform and, with modifications introduced in recent years, can be used to check up to 10 000 recombinants per experiment, allowing large genomic libraries to be screened in a reasonably short time. Nevertheless, the requirement for a probe that is at least partly complementary to the desired gene sometimes makes it impossible to use hybridization in clone identification. On these occasions a different strategy will be needed.

The main alternative to hybridization probing is **immunological screening**. The distinction is that, whereas with hybridization probing the cloned DNA fragment is itself directly identified, an immunological method will detect the protein coded by the cloned gene. Immunological techniques therefore presuppose that the cloned gene is being expressed, so that the protein is being made, and that this protein is not normally present in the host cells.

(a) Antibodies bind to foreign molecules

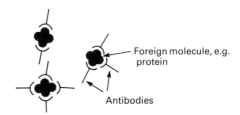

(b) Antibody purification

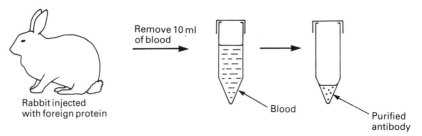

Figure 8.17 Antibodies. (a) Antibodies in the bloodstream bind to foreign molecules and help degrade them. (b) Purified antibodies can be obtained from a small volume of blood taken from a rabbit injected with the foreign protein.

(a) Antibodies are required for immunological detection methods

If a purified sample of a protein is injected into the bloodstream of a rabbit, the immune system of the animal will synthesize antibodies that bind to and help degrade the foreign molecule (Figure 8.17a). This is of course a version of the natural defence mechanism that the animal uses to deal with invasion by bacteria, viruses and other infective agents.

Once challenged with a protein, the levels of antibody present in the animal's bloodstream will remain high enough over the next few days for substantial quantities to be purified. It is not necessary to kill the rabbit, because as little as 10 ml of blood will provide a considerable amount of

(a) Immunoscreening

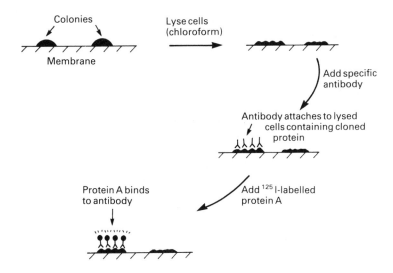

(b) The resulting autoradiograph

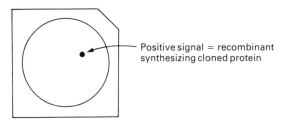

Figure 8.18 Using a purified antibody to detect protein in recombinant colonies.

antibody (Figure 8.17b). This purified antibody will bind only to the protein with which the animal was originally challenged.

(b) Using a purified antibody to detect protein in recombinant colonies

There are several versions of immunological screening, but the most useful method is a direct counterpart of colony hybridization probing. Recombinant colonies are transferred to a polyvinyl membrane, the cells are lysed, and a solution containing the specific antibody is added (Figure 8.18a). Either the antibody itself will be labelled, or the membrane will subsequently be washed with a solution of labelled **Protein A**, a bacterial protein that specifically binds to the immunoglobulins that antibodies are made of (Figure 8.18b). The label can be a radioactive one, in which case the colonies that bind the label will be detected by autoradiography, or non-radioactive labels resulting in a fluorescent or chemiluminescent signal can be used.

(c) The problem of gene expression

Immunological screening obviously depends on the cloned gene being expressed so that the protein translation product is present in the recombinant cells. However, as will be discussed in greater detail in Chapter 11, a gene from one organism is often not expressed in a different organism. In particular, it is very unlikely that a cloned animal or plant gene (with the exception of chloroplast genes) will be expressed in *E. coli* cells.

This problem can be circumvented by using a special type of vector, called an **expression vector** (p. 228), designed specifically to promote expression of the cloned gene in a bacterial host. Immunological screening of recombinant *E. coli* colonies carrying animal genes cloned into expression vectors has in fact been very useful in obtaining genes for several important hormones.

FURTHER READING

Gubler, U. and Hoffman, B. J. (1983) A simple and very efficient method for generating cDNA libraries. *Gene*, **25**, 263–9.

Hames, B. D. and Higgins, S. J. (1985) *Nucleic Acid Hybridisation: A Practical Approach.* IRL, Oxford, UK.

Grunstein, M. and Hogness, D. S. (1975) Colony hybridization: a method for the isolation of cloned cDNAs that contain a specific gene. *Proceedings of the National Academy of Sciences, USA*, **72**, 3961–5.

Benton, W. D. and Davis, R. W. (1977) Screening λgt recombinant clones by hybridization to single plaques *in situ*. *Science*, **196**, 180–2.

Singer, R. H. and Ward, D. C. (1980) Actin gene expression visualized in chicken muscle tissue culture by using *in situ* hybridization with a biotinylated nucleotide analog. *Proceedings of the National Academy of Sciences, USA*, **79**, 7331–5 – a non-radioactive labelling method.

Broome, S. and Gilbert, W. (1978) Immunological screening method to detect specific translation products. *Proceedings of the National Academy of Sciences, USA*, **75**, 2746–9.

Young, R. A. and Davis, R. W. (1983) Efficient isolation of genes by using antibody probes. *Proceedings of the National Academy of Sciences, USA*, **80**, 1194–8.

Studying gene location and structure 9

A carefully designed and skilfully performed cloning experiment will provide a colony or plaque containing copies of the recombinant DNA molecule that carries the gene of interest. In many cases, the next stage in the research project will be to purify the recombinant DNA molecule and obtain from it as much information as possible about the cloned gene.

Many different techniques for studying cloned genes have been developed over the last 15 years and only a representative sample of the most important ones will be covered in this book. In this chapter the emphasis is placed on how to obtain information on the location of a cloned gene on a DNA molecule, and how to study the structure of the cloned gene. In the next chapter methods for studying gene expression will be described. Finally, in Part Three, we will look at how these recombinant DNA techniques are being used in the real world of research and biotechnology.

9.1 HOW TO STUDY THE LOCATION OF A CLONED GENE

Techniques are available for determining the precise location of a cloned gene on the DNA molecule on which it normally resides. The exact nature of the procedure that is used depends on the size of the DNA molecule involved, with the techniques applicable for small molecules, such as plasmids and phage chromosomes, being different from those used for gene location on the large DNA molecules contained in eukaryotic chromosomes.

9.1.1 LOCATING THE POSITION OF A CLONED GENE ON A SMALL DNA MOLECULE

Consider again the example used in the previous chapter where the kanamycin resistance gene from R6.5 was cloned as an *Eco*RI fragment carried by pBR322 (p. 156). Now the clone is available it would be useful to know on which of the 13 R6.5 *Eco*RI fragments the gene lies; this information

(a) Electrophorese EcoRI-restricted R6.5 DNA

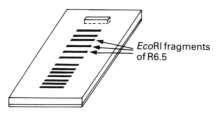

EcoRI fragments of R6.5

(b) Southern transfer

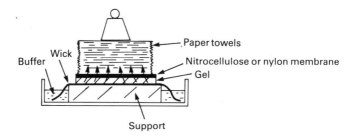

Paper towels

Wick

Buffer

Nitrocellulose or nylon membrane

Gel

Support

(c) Result of hybridization probing

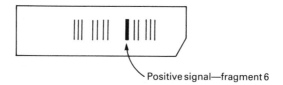

Positive signal—fragment 6

(d) Locate the fragment on the R6.5 restriction map

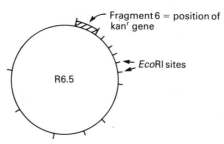

Fragment 6 = position of kan^r gene

EcoRI sites

R6.5

Figure 9.1 Southern hybridization.

will allow the gene to be placed on the R6.5 restriction map and to be positioned relative to other genes.

First, an *Eco*RI restriction digest of R6.5 must be electrophoresed in an agarose gel so that the individual fragments can be seen (Figure 9.1a). One of these fragments will be the same as that inserted into the recombinant pBR322 molecule that carries the kanamycin resistance gene. The aim is therefore to label the recombinant molecule and use it to probe the restriction digest. This can be attempted while the restriction fragments are still contained in the electrophoresis gel, but the results are usually not very good, as the gel matrix causes a lot of spurious background hybridization that obscures the specific hybridization signal. Instead the DNA bands in the agarose gel are transferred to a nitrocellulose or nylon membrane, providing a much 'cleaner' environment for the hybridization experiment.

Transfer of DNA bands from an agarose gel to a membrane makes use of the technique perfected in 1975 by Professor E. M. Southern and referred to as **Southern transfer**. The membrane is placed on the gel, and buffer allowed to soak through, carrying the DNA from the gel to the membrane where the DNA is bound. Sophisticated pieces of apparatus can be purchased to assist this process, but many molecular biologists prefer a homemade set-up incorporating a lot of paper towels and considerable balancing skills (Figure 9.1b). The same method can also be used for the transfer of RNA molecules (**'Northern' transfer**) or proteins (**'Western' transfer**). So far no one has come up with 'Eastern' transfers.

Southern transfer results in a membrane that carries a replica of the DNA bands from the agarose gel. If the labelled probe is now applied, hybridization will occur and autoradiography (or the equivalent detection system for a non-radioactive probe) will show which restriction fragment contains the cloned gene (Figure 9.1c). It will then be possible to position the kanamycin resistance gene on the R6.5 restriction map (Figure 9.1d).

Southern hybridization can also be used to locate the exact position of a cloned gene within a recombinant DNA molecule. This is important as often a cloned DNA fragment will be relatively large (40 kb for a cosmid vector) whereas the gene of interest, contained somewhere in the cloned fragment, may be less than 1 kb in size. Also the cloned fragment may carry a number of genes in addition to the one under study. The strategies described in Chapter 8 for identifying a clone from a genomic library must therefore be followed up by Southern analysis of the recombinant DNA molecule to locate the precise position within the cloned DNA fragment of the gene being sought (Figure 9.2).

9.1.2 LOCATING THE POSITION OF A CLONED GENE ON A LARGE DNA MOLECULE

Southern hybridization is feasible only if a restriction map can be worked out for the DNA molecule being studied. This means that the procedure is appropriate for most plasmids, bacteriophages and viruses, but cannot be

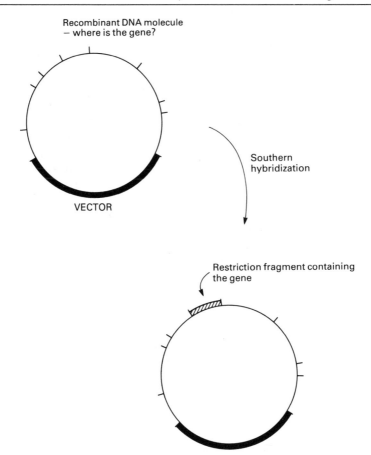

Figure 9.2 Southern hybridization can be used to locate the position of a cloned gene within a recombinant DNA molecule.

used to locate cloned genes on larger DNA molecules. Restriction mapping becomes very complicated with molecules more than about 250 kb in size, as can be appreciated by referring back to Figure 4.18. This example of restriction mapping is relatively straightforward as the λ molecule is not very big. Imagine how much more complicated the analysis would be if there were five times as many restriction sites.

Other techniques have to be used to locate the positions of cloned eukaryotic genes on chromosomal DNA molecules. Of course if the cloned gene has previously been studied by classical genetic methods then its location will probably be known, having already been established by genetic mapping techniques. However, in many cases the most interesting cloned genes are ones that have not been identified by classical genetics. How then can such a gene be mapped?

(a) Separating chromosomes by gel electrophoresis

The first question to ask is: which chromosome carries the gene of interest? For some organisms this can be answered by a special type of Southern hybridization, involving not restriction fragments but intact chromosomal DNA molecules, separated by a novel type of gel electrophoresis.

In conventional gel electrophoresis, as described on p. 66, the electric field is orientated along the length of the gel and the DNA molecules migrate in a straight line towards the positive pole (Figure 9.3a). Different-sized molecules can be separated because of the different rates at which they are able to migrate through the network of pores that make up the gel. However, only molecules within a certain size range can be separated in this way, because the difference in migration rate becomes increasingly small for larger molecules (Figure 9.3b). In practice, molecules larger than 60 kb cannot be resolved efficiently by standard gel electrophoresis.

The limitations of standard gel electrophoresis can be overcome if a more complex electric field is used. Several different systems have been designed

(a) Conventional agarose gel electrophoresis

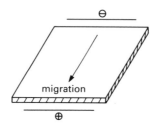

(b) The influence of DNA size on migration rate

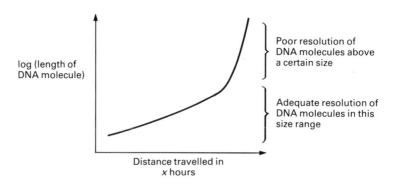

Figure 9.3 Conventional gel electrophoresis and its limitations.

recently but the principle is best illustrated by **orthogonal field alternation gel electrophoresis (OFAGE)** (Figure 9.4a). Instead of being applied directly along the length of the gel, the electric field now alternates rapidly between two pairs of electrodes, each pair set at a 45° angle. The result is a pulsed field, with the DNA molecules in the gel having continually to change direction in accordance with the pulses.

As the two fields alternate in a regular fashion the net movement of the DNA molecules in the gel is still from one end to the other, in more or less a straight line (Figure 9.4a). However, with every change in field direction each DNA molecule has to realign through 90° before its migration can continue. This is in fact the key point, because a short molecule can realign faster than a long one, allowing the short molecule to progress towards the bottom of the gel more quickly. This added dimension increases the resolving power of the gel quite dramatically, so that molecules up to several thousand kb can be separated. This size range includes the chromosomal molecules of many eukaryotes, such as yeast and filamentous fungi,

(a) OFAGE

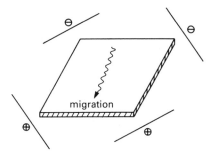

(b) Separation of yeast chromosomes by OFAGE

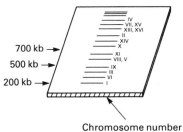

Chromosome number

Figure 9.4 Orthogonal field alternation gel electrophoresis (OFAGE).

meaning that gels showing the chromosomes of these organisms can be obtained (Figure 9.4b).

OFAGE and related techniques are important for a number of reasons. For example, the DNA from individual chromosomes can be purified from the gel, enabling a series of chromosomal gene libraries to be prepared. Each of these libraries, containing the genes from just one chromosome, will be substantially smaller and easier to handle than a complete genomic library. In addition, the chromosomal DNA molecules from an OFAGE gel can be immobilized on a nitrocellulose or nylon membrane by Southern transfer and studied by hybridization analysis. In this way the chromosome that carries a cloned gene can be identified.

(b) *In situ* hybridization to visualize the position of a cloned gene on a eukaryotic chromosome

OFAGE and other non-conventional gel electrophoresis techniques are at present limited to lower eukaryotes whose chromosomes are relatively small. The chromosomal DNA molecules of higher organisms such as man are 100 000 kb and more in length, which is still some way beyond the capability of the current technology. Furthermore, hybridization analysis of chromosomal DNA molecules tells you only the chromosome your cloned gene is located on, and not the exact position of the gene on that chromosome. The technique called *in situ* **hybridization** is more appropriate to higher eukaryotes and can provide real positional information.

With many higher organisms, light microscopy can be used to observe chromosomes in cells that are in the process of division (Figure 9.5a). Individual chromosomes can be recognized by their shape and by the banding pattern produced by various types of stain. *In situ* hybridization provides a direct visual localization of a cloned gene on the light microscopic image of a chromosome.

Cells are treated with a fixative, attached to a glass slide and then incubated with ribonuclease and sodium hydroxide to degrade RNA and denature the DNA molecules. Base pairing between the individual polynucleotide strands is broken down, and the chromosomes unpack to a certain extent, exposing segments of DNA normally enclosed within their structure (Figure 9.5b). A sample of the cloned gene is then labelled and applied to the chromosome preparation. Hybridization occurs between the cloned gene and its chromosomal copy, resulting in a dark spot on an autoradiograph. The position of this spot will indicate the location of the cloned gene on the chromosome.

In situ hybridization is not a very sensitive technique, and often non-specific background hybridization will produce spurious spots on the autoradiograph. Nevertheless, several genes have been placed on the human cytogenetic map using this procedure, and future refinements will undoubtedly increase its value.

(a) Human chromosomes at metaphase

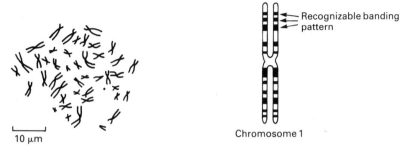

10 μm

Recognizable banding pattern

Chromosome 1

(b) *In situ* hybridization

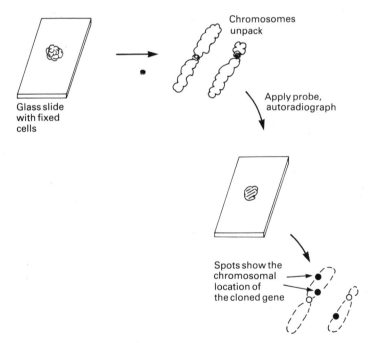

Glass slide with fixed cells

Chromosomes unpack

Apply probe, autoradiograph

Spots show the chromosomal location of the cloned gene

Figure 9.5 Chromosomes and *in situ* hybridization.

(c) Walking along a chromosome from one gene to another

So far we have looked at ways of finding the position on a chromosome of a gene that has already been cloned. Quite frequently though the reverse problem arises: the position of the gene is known, but no probe is available for it, so the relevant clone cannot be isolated from a genomic library. **Chromosome walking** allows the desired clone to be identified, but only so long as a clone is already available for a second gene, one that is known to have a chromosomal location near to the desired gene.

Two clone libraries are needed, each prepared with a different restriction endonuclease, say *Eco*RI for one and *Bam*HI for the other. The fragments carried by the clones in the two libraries will overlap (Figure 9.6a). To begin the chromosome walk the clone containing the known gene is taken from library A and used to probe library B (Figure 9.6b). One or more clones from library B will give positive hybridization signals, showing that the fragments carried by these clones overlap with the fragment carried by the probe. One of these clones from library B is now used to probe library A (Figure 9.6c). The original clone, and possibly some others, will hybridize. The cycle is repeated several times, gradually building up a partial restriction map of the overlapping fragments, until the gene being sought is reached (Figure 9.6d).

Chromosome walking is a difficult and tedious technique. To date the longest walks that have been achieved are only 200 to 250 kb in length. However, the procedure has had a number of outstanding successes, notably in locating the human cystic fibrosis gene. In Chapter 12 we will look at exactly how chromosome walking was used in this project.

9.2 DNA SEQUENCING – WORKING OUT THE STRUCTURE OF A GENE

Probably the most important technique available to the molecular biologist is DNA sequencing, by which the precise order of nucleotides in a piece of DNA can be determined. DNA sequencing methods have been around for several years, but only since the late 1970s has rapid and efficient sequencing been possible. Two different techniques were developed almost simul-taneously – the chain termination method by F. Sanger and A. R. Coulson in the UK, and the chemical degradation method by A. Maxam and W. Gilbert in the USA. The two techniques are radically different but equally valuable. Both allow DNA sequences of several kb in length to be determined in the minimum of time. The DNA sequence is now the first and most basic type of information to be obtained about a cloned gene.

9.2.1 THE SANGER–COULSON METHOD – CHAIN-TERMINATING NUCLEOTIDES

The chain termination method requires single-stranded DNA and so the molecule to be sequenced is usually cloned into an M13 vector. This is

(a) Two clone libraries

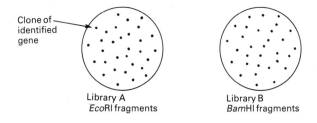

Clone of identified gene

Library A
EcoRI fragments

Library B
BamHI fragments

(b) Probe B with clone 1

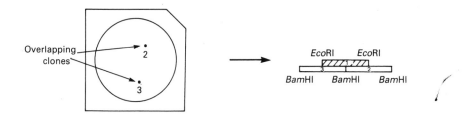

Overlapping clones

2

3

EcoRI   EcoRI

BamHI   BamHI   BamHI

(c) Probe A with clone 2

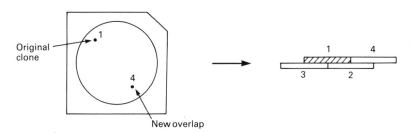

Original clone

1

4

New overlap

1   4

3   2

(d) Continue

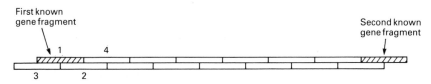

First known gene fragment

Second known gene fragment

1   4

3   2

Figure 9.6 Chromosome walking.

because chain termination sequencing involves the enzymatic synthesis of a second strand of DNA, complementary to an existing template.

(a) The primer

The first step in a chain termination sequencing experiment is to anneal a short oligonucleotide primer on to the recombinant M13 molecule (Figure 9.7a). This primer will act as the starting point for the complementary strand synthesis reaction carried out by the Klenow fragment of DNA polymerase I, or a related enzyme such as T7 DNA polymerase. Remember that these enzymes need a double-stranded region from which to begin strand synthesis (p. 54). The primer anneals to the vector at a position adjacent to the polylinker.

(b) Synthesis of the complementary strand

The strand synthesis reaction is started by adding the enzyme plus each of the four deoxynucleotides (dATP, dTTP, dGTP, dCTP). In addition a single modified nucleotide is also included in the reaction mixture. This is a **dideoxynucleotide** (e.g. dideoxyATP) which can be incorporated into the growing polynucleotide strand just as efficiently as the normal nucleotide, but which blocks further strand synthesis. This is because the dideoxy-nucleotide lacks the hydroxyl group at the 3' position of the sugar component (Figure 9.7b). This group is needed for the next nucleotide to be attached; chain termination therefore occurs whenever a dideoxynucleotide is incorporated by the enzyme.

If dideoxylATP is added to the reaction mix, then termination will occur at positions opposite thymidines in the template (Figure 9.7c). But termination will not always occur at the first T as normal dATP is also present and may be incorporated instead of the dideoxynucleotide. The ratio of dATP to dideoxyATP is such that an individual strand may be polymerized for a considerable distance before a dideoxyATP molecule is added. The result is that a family of new strands is obtained, all of different lengths, but each ending in dideoxyATP.

(c) Four separate reactions result in four families of terminated strands

The strand synthesis reaction is carried out four times in parallel. As well as the reaction with dideoxyATP, there will be one with dideoxyTTP, one with dideoxyGTP, and one with dideoxyCTP. The result will be four distinct families of newly synthesized polynucleotides, one family containing strands all ending in dideoxyATP, one of strands ending in dideoxyTTP, etc.

The next step is to separate the components of each family so the lengths

(a) Anneal the primer

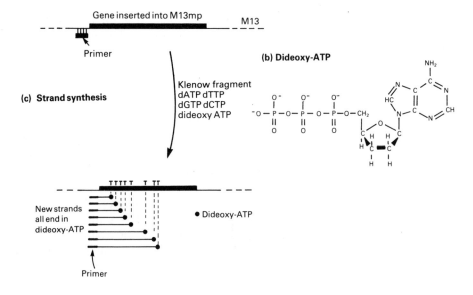

(b) Dideoxy-ATP

(c) Strand synthesis

Klenow fragment
dATP dTTP
dGTP dCTP
dideoxy ATP

New strands
all end in
dideoxy-ATP

● Dideoxy-ATP

Primer

(d) Resulting autoradiograph

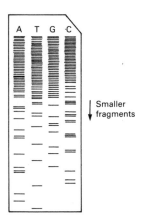

Smaller fragments

Figure 9.7 Chain-termination DNA sequencing.

of each strand can be determined. This can be achieved by gel electro-phoresis, although the conditions have to be carefully controlled as it will be necessary to separate strands that differ in length by just one nucleotide. In practice, the electrophoresis is carried out in very thin polyacrylamide gels (less than 0.5 mm thick). The gels contain urea, which denatures the DNA so the newly synthesized strands dissociate from the templates. In addition, the electrophoresis is carried out at a high voltage, so the gel heats up to 60°C and above, making sure the strands do not reassociate in any way.

Each band in the gel will contain only a small amount of DNA, so autoradiography has to be used to visualize the results (Figure 9.7d). The label is introduced into the new strands by including a radioactive deoxynucleotide (^{32}P- or ^{35}S-dATP) in the reaction mixture for the strand synthesis step earlier in the experiment.

(d) Reading the DNA sequence from the autoradiograph

Reading the sequence is very easy (Figure 9.8). First the band that has moved the furthest is located. This will represent the smallest piece of DNA, the strand terminated by incorporation of the dideoxynucleotide at the first position in the template. The track in which this band occurs is noted. Let us say it is track A; the first nucleotide in the sequence is therefore A.

The next most mobile band will correspond to a DNA molecule one nucleotide longer than the first. The track is noted, T in the example shown in Figure 9.8; the second nucleotide is therefore T and the sequence so far is AT.

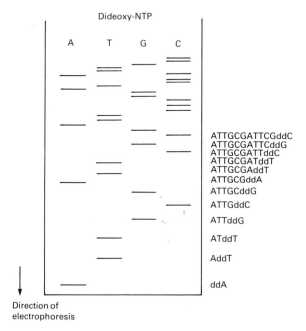

Figure 9.8 Interpreting the autoradiograph produced by a chain-termination sequencing experiment. Each track contains the fragments produced by strand synthesis in the presence of one of the four dideoxyNTPs. The sequence is read by identifying the track that each fragment lies in, starting with the one that has moved the furthest, and gradually progressing up through the autoradiograph.

The process is continued along the autoradiograph until the individual bands become so bunched up that they cannot be separated from one another. Generally it is possible to read a sequence of about 400 nucleotides from one autoradiograph.

9.2.2 THE MAXAM–GILBERT METHOD – CHEMICAL DEGRADATION OF DNA

There are only a few similarities between the Sanger–Coulson and Maxam–Gilbert methods of DNA sequencing. The Maxam–Gilbert method requires double-stranded DNA fragments, so cloning into an M13 vector is not an essential first step. Neither is a primer needed, because the basis of the Maxam–Gilbert technique is not synthesis of a new strand, but cleavage of the existing DNA molecule using chemical reagents that act specifically at a particular nucleotide.

There are several variations of the Maxam–Gilbert method, differing in the exact nature of the cleavage reagents that are used. Most of these are very toxic – chemicals that cleave DNA molecules in the test-tube will do the same in the body, and great care must be taken when using them.

(a) Performing a Maxam–Gilbert sequencing experiment

The double-stranded DNA fragment to be sequenced is first labelled by attaching a radioactive phosphorus group to the 5' end of each strand (Figure 9.9a). Dimethyl sulphoxide is then added and the labelled DNA sample heated to 90°C. This results in breakdown of the base pairing and dissociation of the DNA molecule into its two component strands. The two strands are separated from one another by gel electrophoresis (Figure 9.9b), which works on the basis that one of the strands will contain more purine nucleotides than the other and will therefore be slightly heavier. The heavy strand is purified from the gel and divided into five samples, each of which is treated with one of the cleavage reagents. In fact, the first set of reagents to be added cause a chemical modification in the nucleotides they are specific for, making the strand susceptible to cleavage at that nucleotide when an additional chemical, piperidine, is added (Figure 9.9c). The modification and cleavage reactions are carried out under conditions that result in only one breakage per strand.

Some of the cleaved fragments retain the ^{32}P label at their 5' ends. After electrophoresis, using the same special conditions as for chain termination sequencing, the bands visualized by autoradiography will represent these labelled fragments. The nucleotide sequence can now be read from the autoradiograph exactly as for a chain termination experiment (Figure 9.9d).

9.2.3 BUILDING UP A LONG DNA SEQUENCE

A single DNA sequencing experiment, whether using the Sanger–Coulson or the Maxam–Gilbert method, will produce only about 400 nucleotides of

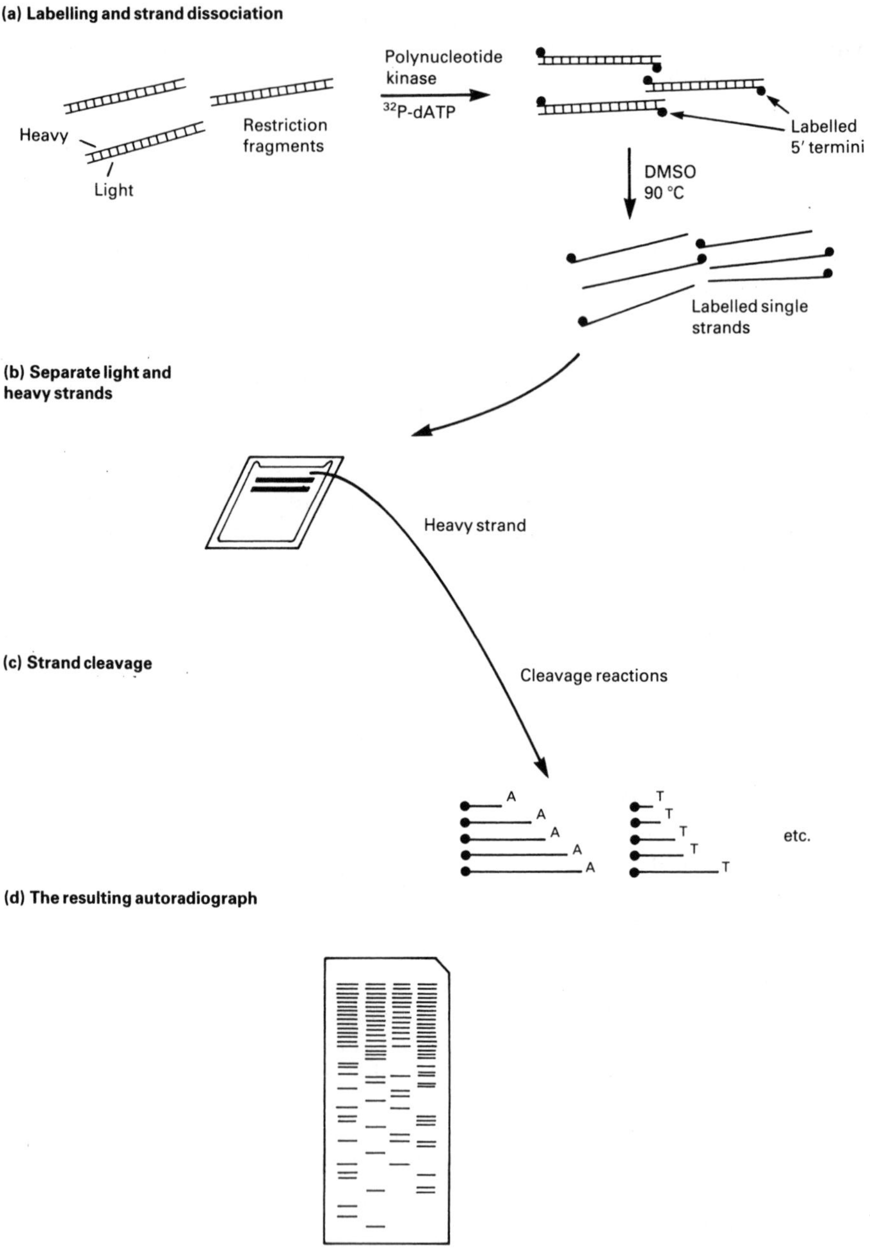
(a) Labelling and strand dissociation
Heavy
Light
Restriction fragments
Polynucleotide kinase
^{32}P-dATP
Labelled 5′ termini
DMSO 90 °C
Labelled single strands
(b) Separate light and heavy strands
Heavy strand
(c) Strand cleavage
Cleavage reactions
A
A
A
A
A
T
T
T
T
T
etc.
(d) The resulting autoradiograph

Figure 9.9 DNA sequencing by the chemical degradation method.

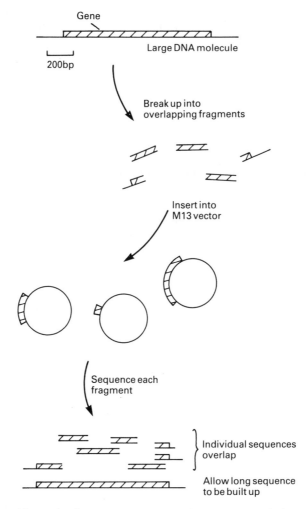

Figure 9.10 Building up a long DNA sequence from a series of short overlapping ones.

sequence. But most genes will be much longer than this; how can a sequence of several kb be obtained?

The answer is to perform DNA sequencing experiments with a lot of different fragments, all derived from a single larger DNA molecule (Figure 9.10). These fragments should overlap, so the individual DNA sequences will themselves overlap. The overlaps can be located, either by eye or using a computer, and the master sequence gradually built up.

There are several ways of producing the overlapping fragments. For instance, the DNA molecule could be cleaved with two different restriction endonucleases, producing one set of fragments with say *Sau*3A and another with *Alu*I (Figure 9.11). This is the traditional method of producing overlapping sequences but suffers from the drawback that the restriction sites may be inconveniently placed and individual fragments may be too long to be completely sequenced. Often four or five different restriction endonucleases will have to be used to clear up all the gaps in the master sequence. However, this is still the easiest and most foolproof method of obtaining a fully contiguous sequence.

9.2.4 THE POWER OF DNA SEQUENCING

The first DNA molecule to be completely sequenced was the 5386 bp chromosome of the bacteriophage ΦX174, which was completed in 1975. This was quickly followed by sequences for SV40 virus (5243 bp) in 1977 and pBR322 (4363 bp) in 1978. Gradually sequencing was applied to larger molecules. Professor Sanger's group published the sequence of the human mitochondrial chromosome (16.5 kb) in 1981 and of bacteriophage λ (49 kb) in 1982. Nowadays sequences of 5 to 10 kb are routine and most research laboratories have the necessary expertise to generate this amount of information.

The pioneering projects today are the massive genome sequencing initiatives, each aimed at obtaining the nucleotide sequence of the entire genome of a particular organism. The human genome project is beginning to get underway with the objective of generating some 3 million kb of nucleotide sequence. Smaller but still massive projects are being planned for the yeast *Saccharomyces cerevisiae* and the plant *Arabidopsis thaliana*.

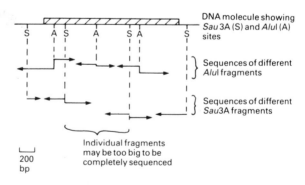

Figure 9.11 Building up a long DNA sequence by determining the sequences of overlapping restriction fragments.

9.3 OBTAINING LESS PRECISE INFORMATION ON GENE STRUCTURE

Often a molecular biologist will not be concerned with the complete nucleotide sequence of a gene, but will nonetheless want to obtain a certain amount of information on its structure. An important instance occurs in screening for genetic diseases (p. 258), where less direct methods than sequencing are used to determine if the DNA from an individual contains a correct or an altered version of a key gene.

9.3.1 RESTRICTION FRAGMENT LENGTH POLYMORPHISM ANALYSIS

An alternative to complete DNA sequencing is provided by restriction analysis. If two DNA molecules are essentially the same, but nonetheless have one or more small differences in their nucleotide sequence, then the fact they are not absolutely identical may be apparent from a comparison of their restriction maps. Three possibilities are illustrated in Figure 9.12, each

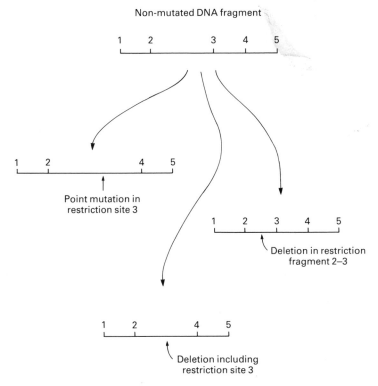

Figure 9.12 Three mutations and the effects they will have on the restriction map of a DNA molecule.

of which will result in a **restriction fragment length polymorphism (RFLP)** that would be apparent if the restriction fragments concerned could be visualized by agarose gel electrophoresis.

The problem is that if we are dealing with a higher organism then the individual fragments will not be seen after electrophoresis. Restriction of human DNA with *Eco*RI, for instance, will result in approximately 700 000 different fragments, producing a smear in an agarose gel from which it will be absolutely impossible to distinguish a single altered band.

In practice, a modified strategy based on Southern hybridization is used. The restriction digest is electrophoresed in the normal way but, rather than being visualized by ethidium bromide staining, the smear of fragments is transferred to a nitrocellulose or nylon membrane by Southern blotting (Figure 9.13). Hybridization analysis is then carried out, using as the probe a clone that spans the entire region of interest. This fragment need not be from the DNA preparation being studied, because there will be substantial

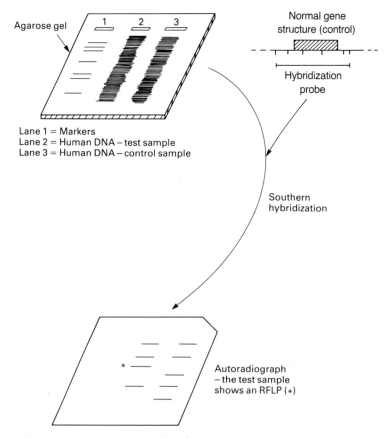

Figure 9.13 Examining a sample of human DNA for a restriction fragment length polymorphism.

hybridization in spite of the small difference in nucleotide sequence. A single probe can therefore be used to analyse many different DNA samples. The probe will hybridize only to the restriction fragments in the relevant region and the sizes of these fragments will appear on the autoradiograph. These sizes can be compared with the pattern for the unmutated gene and an RFLP, if present, will be clearly visible.

9.3.2 THE POLYMERASE CHAIN REACTION PROVIDES AN ALTERNATIVE APPROACH TO RFLP ANALYSIS

A modified form of RFLP analysis can be carried out by the **polymerase chain reaction (PCR)**, a new technique that has applications in many areas of recombinant DNA research.

PCR results in the selective amplification of a chosen region of a DNA molecule (Figure 9.14). Any region of any DNA molecule can be chosen, so long as the sequences at the borders of the region are known. This is because in order to carry out DNA amplification by PCR, two short oligonucleotides must be annealed to the DNA molecule, one to each strand of the double helix (Figure 9.15a). These oligonucleotides delimit the region that will be amplified.

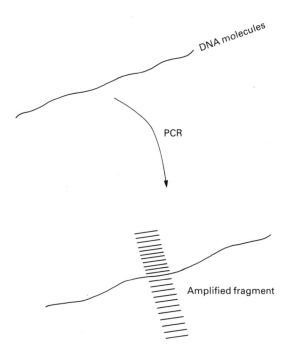

Figure 9.14 Amplification of a specific region of a DNA molecule by the polymerase chain reaction (PCR).

(a) Anneal primers to denatured DNA

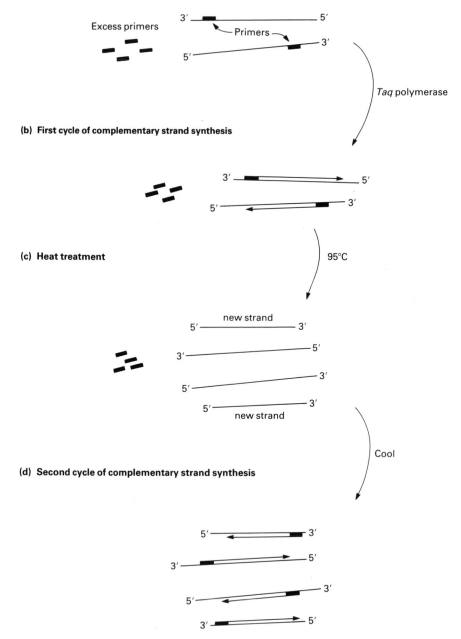

(b) First cycle of complementary strand synthesis

(c) Heat treatment

(d) Second cycle of complementary strand synthesis

Figure 9.15 Polymerase chain reaction (PCR).

Amplification is carried out by the DNA polymerase I enzyme from *Thermus aquaticus*, the same bacterium that produces the restriction enzyme *Taq*I (p. 65). This organism lives at high temperatures in hot springs, and many of its enzymes, including the *Taq* polymerase, are thermostable, meaning that they are resistant to denaturation by heat treatment. As will be apparent in a moment, the thermostability of the *Taq* polymerase is the critical feature that has allowed PCR to be developed.

To begin a PCR amplification, the enzyme is added to the primed template DNA and incubated so that it synthesizes new complementary strands (Figure 9.15b). With most types of DNA polymerase this single reaction step is all that could be carried out, but the thermostability of the *Taq* polymerase means that the reaction mixture can now be heated, to 80°C or more, so that the newly synthesized strands detach from the templates (Figure 9.15c). On cooling, more primers anneal at their respective positions (including positions on the newly synthesized strands), and the *Taq* polymerase, unaffected by the heat treatment, carries out a second round of DNA synthesis (Figure 9.15d). The reaction can be continued through 30–40 cycles, with the DNA amplification proceeding in an exponential fashion.

At the end of the experiment the size of the amplified molecule can be determined by gel electrophoresis. Southern hybridization is not necessary as amplification results in enough DNA to produce a band on an agarose gel. The presence of an insertion or deletion in the amplified fragment will be apparent from its size. Alternatively, if the mutation does not produce a size variation great enough to be detected by gel electrophoresis, the fragment produced by PCR amplification can be sequenced to search for the defect. It may not even be necessary to ligate the fragment into a cloning vector before sequencing, as the PCR amplification may produce sufficient DNA for a clear sequence to be obtained.

FURTHER READING

Southern, E. M. (1975) Detection of specific sequences among DNA fragments separated by gel electrophoresis. *Journal of Molecular Biology*, **98**, 503–7 – Southern hybridization.

Carle, G. E. and Olson, M. V. (1985) An electrophoretic karyotype for yeast. *Proceedings of the National Academy of Sciences, USA*, **82**, 3756–60 – an application of OFAGE.

Bender, W., *et al.* (1983) Chromosome walking and jumping to isolate DNA from the *Ace* and *rosy* loci and the bithorax complex in *Drosophila melanogaster*. *Journal of Molecular Biology*, **168**, 17–33.

Sanger, F., *et al.* (1977) DNA sequencing with chain-terminating inhibitors. *Proceedings of the National Academy of Sciences, USA*, **74**, 5463–7.

Maxam, A. and Gilbert, W. (1977) A new method of sequencing DNA. *Proceedings of the National Academy of Sciences, USA*, **74**, 560–4.

DeLisi, C. (1988) The human genome project. *American Scientist*, **76**, 488–93.

White, T. J. *et al.* (1989) The polymerase chain reaction. *Trends in Genetics*, **5**, 185–9.

Studying gene expression

10

All genes have to be expressed in order to function. The first step in expression is transcription of the gene into a complementary RNA strand (Figure 10.1a). For some genes – for example those coding for transfer RNA and ribosomal RNA molecules – the transcript itself is the functionally important molecule. For other genes the transcript is translated into a protein molecule.

To understand how a gene is expressed, the RNA transcript must be studied. In particular, the molecular biologist will want to know whether the transcript is a faithful copy of the gene, or whether segments of the gene are missing from the transcript (Figure 10.1b). These missing pieces are called introns and considerable interest centres on their structure and possible function. In addition to introns, the exact locations of the start and end points of transcription are important. Most transcripts are copies not only of the gene itself, but also of the nucleotide regions either side of it (Figure 10.1c). The signals that determine the start and finish of the transcription process are only partly understood, but their positions must be located if the expression of a cloned gene is to be studied.

In this chapter we will begin by looking at the methods used for **transcript analysis**. These methods can tell you if a cloned gene contains introns and will also help to position the start and end points for transcription on to the nucleotide sequence. Then we will briefly consider a few of the numerous techniques developed in recent years for examining how expression of a cloned gene is regulated. These techniques are important as aberrations in gene regulation underlie many clinical disorders. Finally, we will consider the difficult problem of how to identify the translation product of a cloned gene.

10.1 STUDYING THE TRANSCRIPT OF A CLONED GENE

Most methods of transcript analysis are based on hybridization between the

(a) Genes are expressed by transcription and translation

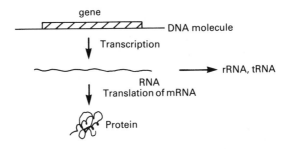

(b) Some genes contain introns

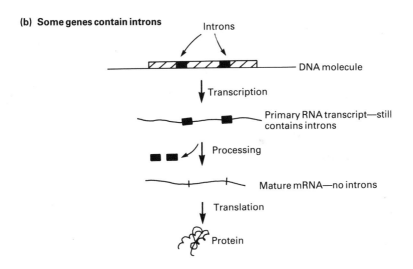

(c) RNA transcripts include regions either side of the gene

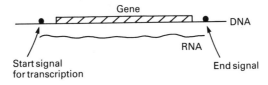

Figure 10.1 Some fundamentals of gene expression.

RNA transcript and a fragment of DNA containing the relevant gene. Nucleic acid hybridization occurs just as readily between complementary DNA and RNA strands as between single-stranded DNA molecules. The resulting DNA–RNA hybrid can be analysed by electron microscopy or with single-strand specific nucleases.

10.1.1 ELECTRON MICROSCOPY OF NUCLEIC ACID MOLECULES

Electron microscopy can be used to observe directly nucleic acid molecules, so long as the polynucleotides are first treated with chemicals that will increase their apparent diameter. Untreated molecules are simply too thin to be seen.

Usually the DNA molecules are mixed with a protein such as cytochrome c which will bind to the polynucleotides, coating the strands in a thick shell. The coated molecules must be stained with uranyl acetate or some other electron-dense material to enhance the appearance of the preparation (Figure 10.2). Quite spectacular views of nucleic acid molecules can be obtained.

In the past, electron microscopy has been used primarily to analyse hybridization between different DNA molecules, but in recent years the technique has become increasingly important in the study of DNA–RNA hybrids. It is particularly useful for determining if a gene contains introns. Consider the appearance of a hybrid between a DNA strand, containing a gene, and its RNA transcript. If the gene contains introns then these regions

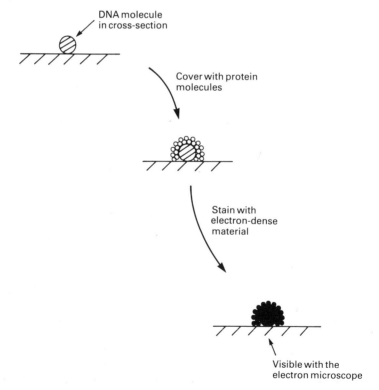

Figure 10.2 Preparing a DNA molecule for observation with the electron microscope.

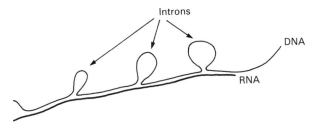

Figure 10.3 The appearance under the electron microscope of a DNA–RNA hybrid formed between a gene containing an intron and its processed transcript.

of the DNA strand will have no homology with the RNA transcript and so will not base pair. Instead they will 'loop out', giving a characteristic appearance when observed with the electron microscope (Figure 10.3). The number and positions of these loops will correspond directly to the number and positions of the introns in the gene. Further information can then be obtained by sequencing the gene and looking for the characteristic features that mark the boundaries of introns. If a cDNA clone is available then its sequence, which will of course lack the introns, can be compared with the gene sequence to locate the introns with precision.

10.1.2 ANALYSIS OF DNA–RNA HYBRIDS BY NUCLEASE TREATMENT

The second method for studying a DNA–RNA hybrid involves a single-strand specific nuclease such as S1 (p. 50). This enzyme degrades single-stranded DNA or RNA polynucleotides, including single-stranded regions at the end of predominantly double-stranded molecules, but has no effect on double-stranded DNA or on DNA–RNA hybrids. If a DNA molecule containing a gene is hybridized to its RNA transcript, and then treated with S1 nuclease, the non-hybridized single-stranded DNA regions at each end of the hybrid will be digested, along with any looped-out introns (Figure 10.4). The result will be a completely double-stranded hybrid. The single-stranded DNA fragments protected from S1 nuclease digestion can be recovered if the DNA strand is degraded by treatment with alkali.

Unfortunately the manipulations shown in Figure 10.4 are not very informative. The sizes of the protected DNA fragments could be measured by gel electrophoresis, but this will not allow their order or relative positions to be determined. However, a few subtle modifications to the technique will allow the precise start and end points of the transcript itself and of any introns it contains to be mapped on to the DNA sequence.

The trick is to select a restriction fragment that spans the end of the gene. In the example shown in Figure 10.5, a *Sau*3A fragment that contains 100 bp of coding region, along with 300 bp of the leader sequence preceding the gene, has been cloned into an M13 vector and obtained as a single-stranded

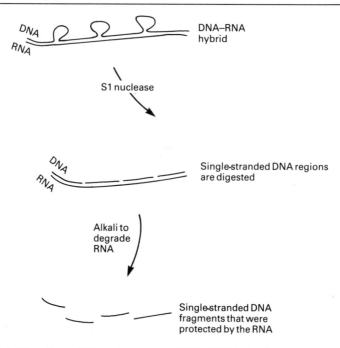

Figure 10.4 The effect of S1 nuclease on a DNA–RNA hybrid.

molecule. A sample of the RNA transcript is added and allowed to anneal to the DNA molecule. The DNA molecule will still primarily be single-stranded but will now have a small region protected by the RNA transcript. All but this protected region is digested by S1 nuclease and the RNA degraded with alkali, leaving a short single-stranded DNA fragment. If these manipulations are examined closely it will become clear that the size of this single-stranded fragment will correspond to the distance between the transcription start point and the righthand *Sau*3A site. The size of the single-stranded fragment is therefore determined by gel electrophoresis and this information is used to locate the transcription start point on the DNA sequence. Exactly the same strategy could locate the end point of transcription and the junction points between introns and exons: the only difference would be the position of the restriction site chosen to delimit one end of the protected single-stranded DNA fragment.

10.1.3 OTHER TECHNIQUES FOR STUDYING RNA TRANSCRIPTS

Electron microscopy and S1 nuclease analysis are not the only methods available for studying RNA transcripts. In recent years a broad range of RNA manipulative techniques have been developed, including reliable methods for direct sequencing of RNA molecules. These methods enable RNA

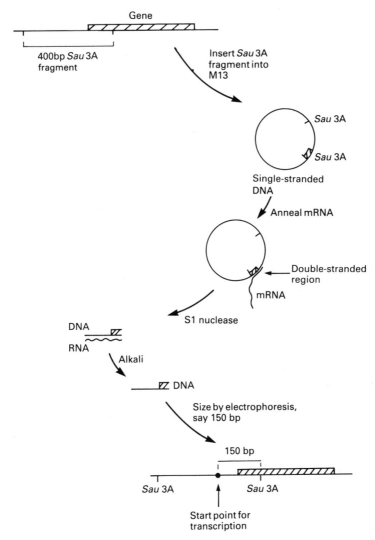

Figure 10.5 Locating a transcription start point by S1 nuclease mapping.

transcription to be studied in ways that have been impossible in the past and, in particular, have resulted in a better understanding of how transcripts are processed. The primary transcript that is copied directly from the gene may undergo a complex series of modification events before the mature mRNA molecule is produced. As well as removal of introns a transcript may be edited by insertion of new nucleotides and by alteration of existing ones. Some RNA molecules may even possess catalytic activity. These exciting discoveries are stimulating the development of new methods for transcript analysis and this is one of many areas of recombinant DNA technology that is progressing rapidly.

10.2 STUDYING THE REGULATION OF GENE EXPRESSION

Few genes are expressed all the time – most are subject to regulation and are switched on only when their gene product is required by the cell. The simplest gene regulation systems are found in bacteria, such as E. coli, which can regulate expression of genes for biosynthetic and metabolic processes, so that gene products that are not needed are not synthesized. For instance the genes coding for the enzymes involved in tryptophan biosynthesis can be switched off when there are abundant amounts of tryptophan in the cell, and switched on again when tryptophan levels drop. Similarly, genes for the utilization of sugars such as lactose are activated only when the relevant sugar is there to be metabolized. In higher organisms gene regulation is more complex because there are many more genes that are subject to control. Differentiation of cells in tissues and organs involves wholesale changes in gene expression patterns, with many of the changes being irreversible. The process of development from fertilized egg cell to adult also involves gene regulation and requires coordination between different cells as well as time-dependent changes in gene expression.

Many of the problems in gene regulation require a classical genetic approach; genetics enables genes that control regulation to be identified, can define the biochemical signals that influence gene expression, and explores the interactions between different genes and gene families. It is for this reason that most of the breakthroughs in understanding development in higher organisms have started with studies of the fruit-fly *Drosophila melanogaster*. Gene cloning complements classical genetics as it provides much more detailed information on the molecular events involved in regulating the expression of a single gene. We now know that a gene subject to regulation has one or more control sequences in its upstream region (Figure 10.6) and that the gene is switched on and off by the attachment of regulatory proteins to these sequences. A regulatory protein may repress

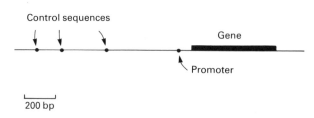

Figure 10.6 Possible positions for control sequences in the region upstream of a gene.

gene expression, in which case the gene is switched off when the protein is bound to the control sequence, or alternatively the protein may have a positive or enhancing role, switching on or increasing expression of the target gene. In this chapter we will examine methods for locating control sequences and determining their exact role in regulating gene expression. The more difficult questions, such as what controls the binding of the regulatory protein to the control sequence, is a problem for genetics and beyond the scope of the simple gene cloner.

10.2.1 IDENTIFYING PROTEIN-BINDING SITES ON A DNA MOLECULE

A control sequence is a region of DNA that can bind a regulatory protein. It should therefore be possible to identify control sequences upstream of a cloned gene by searching the relevant region for protein-binding sites. There are two different approaches. In the first, a DNA fragment that has a protein bound to it is identified by virtue of its increased molecular mass, and in the second the region to which the protein is bound is delineated by its resistance to endonuclease cleavage. In each case the first step is to form a complex between the regulatory protein and the DNA molecule under study. It is very unlikely though that the regulatory protein will be available in pure form, so usually the starting material is an unfractionated extract of nuclear protein (remember that gene regulation occurs in the nucleus).

(a) Gel retardation of DNA–protein complexes

Proteins are quite substantial structures and a protein attached to a DNA molecule will result in a large increase in overall molecular mass. If this increase can be detected then a DNA fragment carrying a protein-binding site will have been identified. In practice a DNA fragment carrying a bound protein is identified by gel electrophoresis, as it will have a lower mobility than the uncomplexed DNA molecule (Figure 10.7). The procedure is referred to as **gel retardation**.

For gel retardation to be of value the region containing the control sequence must be digested with a restriction endonuclease before mixing with the regulatory protein (Figure 10.8). The location of the control sequence is then determined by finding the position on the restriction map of the fragment that binds the regulatory protein and is therefore retarded during gel electrophoresis. The precision with which the control sequence can be located depends on how detailed the restriction map is and how conveniently placed the restriction sites are. A single control sequence may be less than 10 bp in size, so rarely will gel retardation be able to pinpoint it exactly. A more precise technique is therefore needed to delineate the exact position of the protein-binding sequence within the fragment identified by gel retardation.

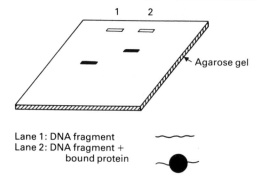

Figure 10.7 A bound protein will lower the mobility of a DNA fragment during gel electrophoresis.

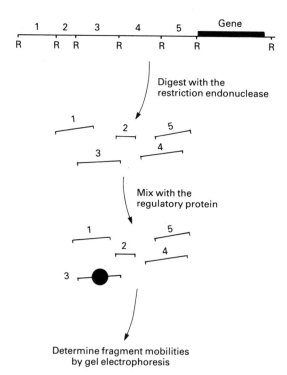

Figure 10.8 Carrying out a gel retardation experiment.

(b) Footprinting with DNase I

The procedure generally called **footprinting** is the desired supplement to gel retardation. Footprinting works on the basis that the interaction with a regulatory protein will protect the DNA in the region of a control sequence from the degradative action of an endonuclease such as DNase I (Figure 10.9). This phenomenon can be used to locate with some precision the protein-binding site on the DNA molecule.

The DNA fragment being studied is first labelled at one end with a radioactive marker, and then complexed with the regulatory protein (Figure 10.10a). DNase I is added, but the amount used is limited so that complete degradation of the DNA fragment does not occur. Instead the aim is to cut each molecule at just a single phosphodiester bond (Figure 10.10b). If the DNA fragment has no protein attached to it then the result of this treatment will be a family of labelled fragments, differing in size by one nucleotide each. After separation on a polyacrylamide gel the family of fragments will appear as a ladder of bands on an autoradiograph (Figure 10.10c). However, the bound protein protects certain phosphodiester bonds from being cut by being cut by DNase I, meaning that the family of fragments will not be complete, as the fragments resulting from cleavage within the control sequence will be absent. Their absence will show up on the autoradiograph as a 'footprint', clearly seen in Figure 10.10c. The position of the control sequence within the DNA molecule can be worked out fairly easily from the size of the fragments on either side of the footprint.

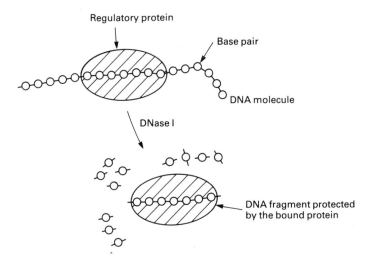

Figure 10.9 A bound protein will protect a region of a DNA molecule from degradation by a nuclease such as DNase I.

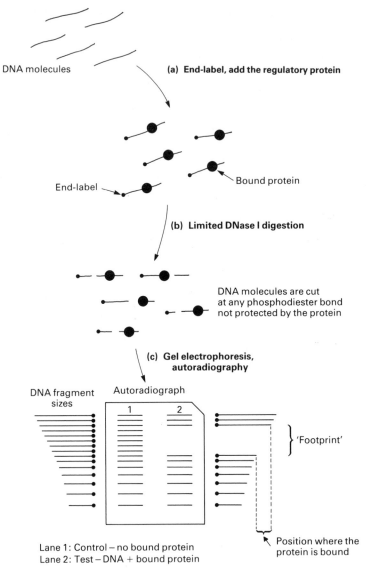

DNA molecules

(a) End-label, add the regulatory protein

End-label

Bound protein

(b) Limited DNase I digestion

DNA molecules are cut
at any phosphodiester bond
not protected by the protein

**(c) Gel electrophoresis,
autoradiography**

DNA fragment
sizes

Autoradiograph

'Footprint'

1   2

Position where the
protein is bound

Lane 1: Control – no bound protein
Lane 2: Test – DNA + bound protein

Figure 10.10 DNase I footprinting.

10.2.2 IDENTIFYING CONTROL SEQUENCES BY DELETION ANALYSIS

Gel retardation and footprinting experiments will locate possible control sequences upstream of a cloned gene but will provide no information on the function of the individual sequences. **Deletion analysis** is a totally different approach that can not only locate control sequences (though only with the precision of gel retardation), but importantly can also indicate the function of each sequence. The technique depends on the assumption that deletion of the control sequence will result in a change in the way in which expression of the cloned gene is regulated (Figure 10.11). For instance, deletion of a sequence that represses expression of a gene should result in that gene being expressed at a higher level. Similarly, tissue-specific control sequences can be identified as their deletion will result in the target gene being expressed in tissues other than the correct one.

(a) Reporter genes

Before carrying out deletion analysis a way must be found to assay the effect of a deletion on expression of the cloned gene. The effect will probably only be observed when the gene is cloned into the species it was originally obtained from: it will be no good assaying for light-regulation of a plant gene if the gene is cloned in a bacterium.

Cloning vectors have now been developed for most organisms (Chapter 7) so cloning the gene under study back into its host should not cause a problem. The difficulty is that the host will almost certainly already possess a copy of the cloned gene. How can changes in the expression pattern of the cloned gene be distinguished from the normal pattern of expression displayed by the host's copy of the gene?

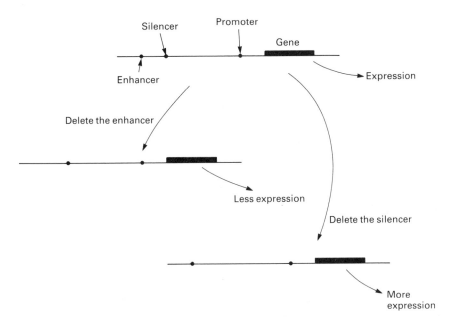

Figure 10.11 The principle behind deletion analysis.

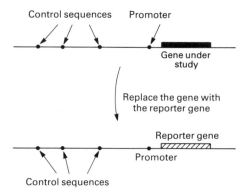

Figure 10.12 A reporter gene.

The answer is to use a **reporter gene**. This is a test gene that is fused to the upstream region of the gene under study (Figure 10.12), replacing the original cloned gene. When cloned into the host organism the expression pattern of the reporter gene will exactly mimic that of the original gene, as the reporter gene will be under the influence of exactly the same control sequences as the original gene.

The reporter gene must be chosen with care. The first criterion is that the reporter gene must code for a phenotype not already displayed by the host organism. The phenotype of the reporter gene must be relatively easy to detect after it has been cloned into the host, and ideally it should be possible to assay the phenotype quantitatively. These criteria have not proved to be difficult to meet and a variety of different reporter genes have been used in studies of gene regulation; a few examples are listed in Table 10.1.

(b) Carrying out a deletion analysis

Once a reporter gene has been chosen and the necessary construction made, carrying out a deletion analysis is fairly straightforward. Deletions can be made in the upstream region of the construct by any one of several strategies, a simple example being shown in Figure 10.13. The effect of the deletion is then assessed by cloning the deleted construct into the host organism and determining the pattern and extent of expression of the reporter gene. An increase in expression will imply that a repressing or silencing sequence has been removed, a decrease will indicate removal of an activator or enhancer, and a change in tissue-specificity (as shown in Figure 10.13) will pinpoint a tissue-responsive control sequence.

The results of a deletion analysis project have to be interpreted very carefully. Complications may arise if a single deletion removes two closely linked control sequences or, as is fairly common, two distinct control sequences cooperate to produce a single response. Despite these potential difficulties deletion analyses, in combination with studies of protein-

Table 10.1 A few examples of reporter genes used in studies of gene regulation in higher organisms

Gene	Gene product	Assay
lacZ	β-Galactosidase	Histochemical test
neo	Neomycin phosphotransferase	Kanamycin resistance
cat	Chloramphenicol acetyltransferase	Chloramphenicol resistance
dhfr	Dihydrofolate reductase	Methotrexate resistance
aphIV	Hygromycin phosphotransferase	Hygromycin resistance
lux	Luciferase	Bioluminescence
uidA	β-Glucuronidase	Histochemical test

All of these genes are obtained from *E. coli*, except for *lux* which has three sources: the luminescent bacteria *Vibrio harveyii* and *V. fischeri*, and the firefly *Photinus pyralis*.

binding sites, have provided important information about how the expression of individual genes is regulated, and have supplemented and extended the more broadly based genetic analyses of differentiation and development.

10.3 IDENTIFYING AND STUDYING THE TRANSLATION PRODUCT OF A CLONED GENE

In the last few years gene cloning has become increasingly useful in the study not only of genes themselves but also of the proteins coded by cloned

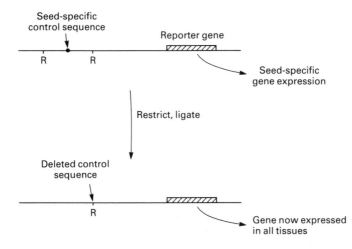

Figure 10.13 Deletion analysis. A reporter gene has been attached to the upstream region of a seed-specific gene from a plant. Removal of the restriction fragment bounded by the sites 'R' deletes the control sequence that mediates seed-specific gene expression, so that the reporter gene is now expressed in all tissues of the plant.

genes. Investigations into protein structure and function have benefited greatly from new techniques that allow mutations to be introduced at specific points in a cloned gene, resulting in directed changes in the structure of the encoded protein.

Before considering these procedures we should first look at the more mundane problem of how to identify the protein coded by a cloned gene. In many cases this analysis will not be necessary as the protein will have been characterized long before the gene cloning experiment is performed. On the other hand, there will be occasions when the translation product of a cloned gene has not been identified. A method for isolating the protein will then be needed.

10.3.1 HRT AND HART CAN IDENTIFY THE TRANSLATION PRODUCT OF A CLONED GENE

Two related techniques, **hybrid-release translation (HRT)** and **hybrid-arrest translation (HART)**, are used to identify the translation product encoded by a cloned gene. Both depend on the ability of purified mRNA to direct synthesis of proteins in **cell-free translation systems**. These are cell extracts, usually prepared from germinating wheat seeds or from rabbit reticulocyte cells (both of which are exceptionally active in protein synthesis) and containing ribosomes, tRNAs and all the other molecules needed for protein synthesis. The mRNA sample is added to the cell-free translation system, along with a mixture of the 20 amino acids found in proteins, one of which is labelled (often ^{35}S-methionine is used). The mRNA molecules will be translated into a mixture of radioactive proteins (Figure 10.14), which can be separated by gel electrophoresis and visualized by autoradiography. Each band represents a single protein coded by one of the mRNA molecules present in the sample.

HRT and HART work best when cDNA clones prepared directly from the mRNA sample are available. For HRT the cDNA is denatured, immobilized on a nitrocellulose membrane, and incubated with the mRNA sample (Figure 10.15). The specific mRNA counterpart of the cDNA will hybridize and remain attached to the membrane. After discarding the unbound molecules, the hybridized mRNA is recovered and translated in a cell-free system. This will produce a pure sample of the protein coded by the cDNA.

HART is slightly different in that the denatured cDNA is added directly to the mRNA sample (Figure 10.16). Hybridization again occurs between the cDNA and its mRNA counterpart but in this case the unbound mRNA is not discarded. Instead the entire sample is translated in the cell-free system. The hybridized mRNA will be unable to direct translation, so all the proteins except the one coded by the cloned gene will be synthesized. The cloned gene's translation product is therefore identified as the protein that is absent from the autoradiograph.

Cell-free translation produces quite small amounts of protein but nevertheless the products of HRT and HART may be sufficient for at least partial

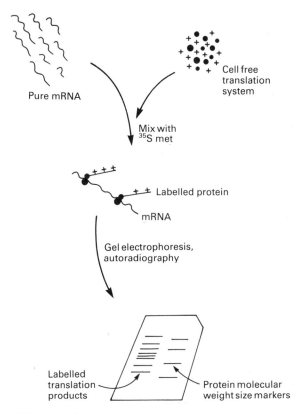

Figure 10.14 Cell-free translation.

characterization of the translation product of the cloned gene. Alternatively the cell-free translation products can be compared with an electrophoretogram of the entire protein complement of the cell from which the gene was obtained. It may then be possible to isolate the required protein in quite large amounts from the total protein extract.

10.3.2 ANALYSIS OF PROTEINS BY *IN VITRO* MUTAGENESIS

Although HRT and HART can identify the translation products of a cloned gene, these techniques will tell us little about the protein itself. The major questions asked by the molecular biologist today centre on the relationship between the structure of a protein and its mode of activity. The best way of tackling these problems would be to induce a mutation in the gene coding for the protein and then to determine what effect the change in amino acid sequence has on the properties of the translation product (Figure 10.17). However, under normal circumstances mutations occur randomly and a

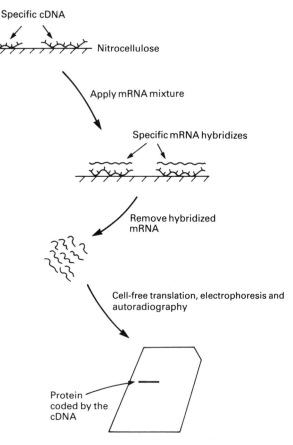

Specific cDNA

Nitrocellulose

Apply mRNA mixture

Specific mRNA hybridizes

Remove hybridized
mRNA

Cell-free translation, electrophoresis and
autoradiography

Protein
coded by the
cDNA

Figure 10.15 Hybrid-release translation.

large number may have to be screened before one that provides useful information is found. Clearly some way of causing directed mutations at a specific point in a gene is needed.

The modern technique of *in vitro* mutagenesis can be used to make these specific alterations in a cloned gene. To be of any use, however, the gene must be transcribed and translated in the host organism so that the mutated protein can be obtained. For an *E. coli* gene cloned in *E. coli* this is no problem; for genes from higher organisms cloned in *E. coli* an expression vector must be used (p. 228).

(a) Different types of *in vitro* mutagenesis techniques

An almost unlimited variety of DNA manipulations can be used to introduce mutations into cloned genes. The following are the simplest.

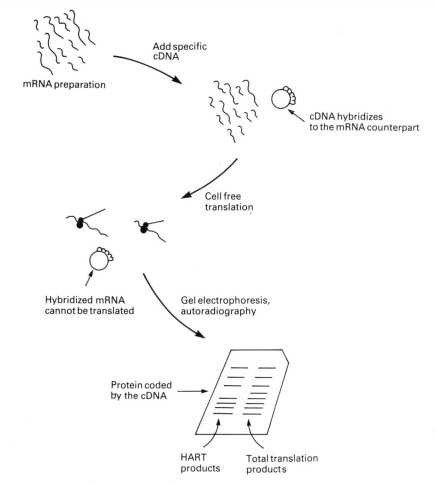

Figure 10.16 Hybrid-arrest translation.

1. A restriction fragment can be deleted (Figure 10.18a).
2. The gene can be opened at a unique restriction site, a few nucleotides removed with a double-strand specific endonuclease such as Bal31 (p. 50), and the gene religated (Figure 10.18b).
3. A short oligonucleotide can be inserted at a restriction site (Figure 10.18c). The sequence of the oligonucleotide can be such that the additional stretch of amino acids inserted into the protein will produce, for example, a new structure such as an α-helix, or will destabilize an existing structure.

Although potentially useful, these manipulations depend on the fortuitous occurrence of a restriction site at the area of interest in the cloned gene.

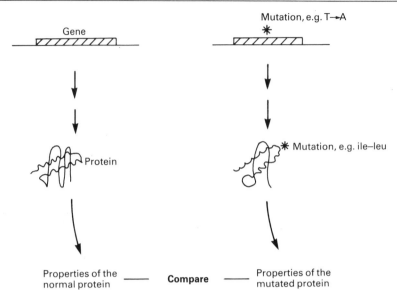

Figure 10.17 A mutation may change the amino acid sequence of a protein, possibly affecting its properties.

Oligonucleotide-directed mutagenesis is a more versatile technique that can introduce a mutation at any point in the gene.

(b) Using an oligonucleotide to introduce a point mutation in a cloned gene

For oligonucleotide-directed mutagenesis the gene must usually be obtained in a single-stranded form and so is generally cloned into an M13 vector. The single-stranded DNA is purified and the region to be mutated identified by DNA sequencing. A short oligonucleotide is then synthesized, complementary to the relevant region, but containing the desired nucleotide alteration (Figure 10.19a). Despite this mismatch the oligonucleotide will anneal to the single-stranded DNA and act as a primer for complementary strand synthesis using the Klenow fragment of DNA polymerase I (Figure 10.19b). This strand-synthesis reaction is continued until an entire new strand is made and the recombinant molecule is completely double-stranded.

After introduction, by transfection, into competent *E. coli* cells, DNA replication will produce numerous copies of the molecule (Figure 10.19c). The semiconservative nature of DNA replication means that half the phage progeny will carry copies of the original, unmutated recombinant M13 molecule, whereas half will contain the altered molecule. The phage produced by the transfected cells are plated on to solid agar so that plaques are produced. Half the plaques should contain the original recombinant

(a) Restriction fragment deletion

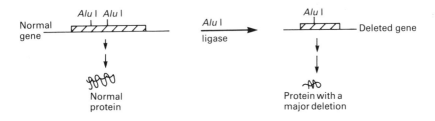

(b) Nucleotide removal at restriction site

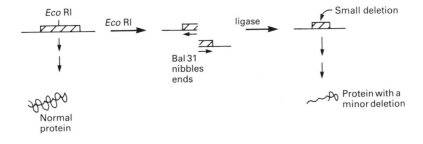

(c) Insertion of an oligonucleotide

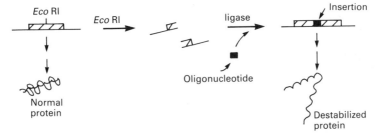

Figure 10.18 Various *in vitro* mutagenesis techniques.

molecule, and half the mutated version. Which are which is determined by plaque hybridization, using the oligonucleotide as the probe, and employing very strict conditions so that only the completely base-paired hybrid is stable.

(c) Studying the effect of the mutation

Cells infected with M13 vectors do not lyse, but instead continue to divide (p. 19). A gene ligated into an M13 vector can therefore be expressed in the

(a) The oligonucleotide

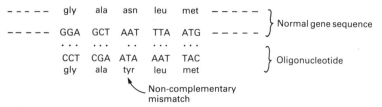

(b) Complementary strand synthesis

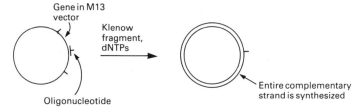

(c) Isolation of mutant E.coli

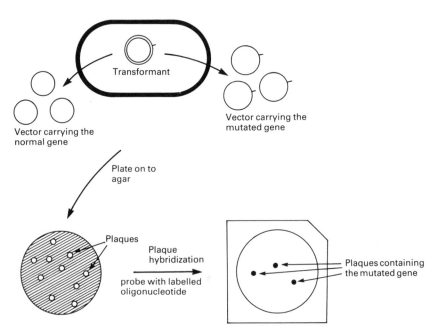

Figure 10.19 One method for oligonucleotide-directed mutagenesis.

host cells resulting in production of recombinant protein. The protein coded by the cloned gene can be purified from the recombinant cells and its properties studied. The effect of a single base-pair mutation on the activity of the protein can therefore be assessed.

(d) The potential of oligonucleotide-directed mutagenesis

This technique has remarkable potential, both for pure research and for applied biotechnology. The biochemist can now ask very specific questions about the way that protein structure affects the action of enzymes. Biochemical analysis can provide an idea of which amino acids in an enzyme participate in substrate binding and in the catalytic reaction. Oligonucleotide-directed mutagenesis can be used to alter the amino acid sequence at the important sites in order to examine in detail the role of individual residues in substrate binding and catalytic activity. Oligonucleotide-directed mutagenesis will undoubtedly provide a boost to our understanding of enzymatic catalytic mechanisms.

In biotechnology, several important proteins (for example insulin and the blood-clotting protein Factor VIII) are now being produced from cloned genes (Chapter 12). Oligonucleotide-directed mutagenesis could be employed to alter the structure of the cloned genes, possibly to result in a more active protein, possibly to remove an unwanted side-effect, possibly to improve the stability of the protein. Protein engineering, through gene cloning and oligonucleotide-directed mutagenesis, promises to become one of the major biological growth areas of the 1990s.

FURTHER READING

Favaloro, J. *et al.* (1980) Transcription maps of polyoma virus-specific RNA: analysis by two-dimensional nuclease S1 gel mapping. *Methods in Enzymology*, **65**, 718–49.

Fried, M. and Crothers, D. M. (1981) Equilibria and kinetics of lac repressor–operator interactions by polyacrylamide gel electrophoresis. *Nucleic Acids Research*, **9**, 6505–25 – gel retardation.

Garnier, M. M. and Revzin, A. (1981) A gel electrophoretic method for quantifying the binding of proteins to specific DNA regions. *Nucleic Acids Research*, **9**, 3047–60 – gel retardation.

Galas, D. J. and Schmitz, A. (1978) DNase footprinting: a simple method for the detection of protein–DNA binding specificity. *Nucleic Acids Research*, **5**, 3157–70.

Paterson, B. M. *et al.* (1977) Structural gene identification and mapping by DNA.mRNA hybrid-arrested cell-free translation. *Proceedings of the National Academy of Sciences, USA*, **74**, 4370–4.

Smith, M. (1985) *In vitro* mutagenesis. *Annual Review of Genetics*, **19**, 423–62.

Part Three

Gene Cloning in Research and Biotechnology

Production of protein from cloned genes 11

Now that we have covered the basic techniques involved in cloning a gene and studying its structure and expression, we can move on to consider how gene cloning is being applied in biological research. This will lead us into the major growth industry of our generation – biotechnology.

Biotechnology is not a new subject, although it has received far more attention during the last few years than it ever has in the past. Biotechnology can be defined as the use of living organisms in industrial or industrial-type processes. According to archaeologists, the British biotechnology industry dates back 4000 years, to the late Neolithic period, when fermentation processes that make use of living yeast cells to produce ale and mead were first introduced into this country. Certainly brewing was well established by the time of the Roman invasion.

During the twentieth century, biotechnology has expanded with the development of a variety of industrial uses for microorganisms. The discovery by Alexander Fleming in 1929 that the mould *Penicillium* synthesizes a potent antibacterial agent led to the use of fungi and bacteria in the large-scale production of antibiotics. At first the microorganisms were grown in large culture vessels from which the antibiotic was purified after the cells had been removed (Figure 11.1a), but more recently this **batch culture** method has been largely supplanted by **continuous culture** techniques, making use of a **fermenter**, from which samples of medium can be continuously drawn off, providing a non-stop supply of the product (Figure 11.1b). This type of process is not limited to antibiotic production and has also been used to obtain large amounts of other compounds produced by microorganisms (Table 11.1).

One of the reasons why biotechnology has received so much attention during the last decade is because of gene cloning. Although many useful products can be obtained from microbial culture, the list in the past has been limited to those compounds naturally synthesized by microorganisms. Many important pharmaceuticals that are produced not by microbes but by higher organisms could not be obtained in this way. This has been changed by the application of gene cloning to biotechnology. The ability to clone

(a) Batch culture

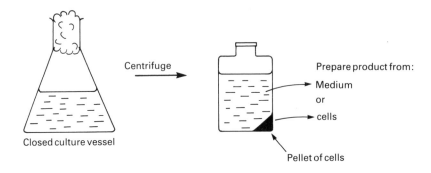

(b) Continuous culture

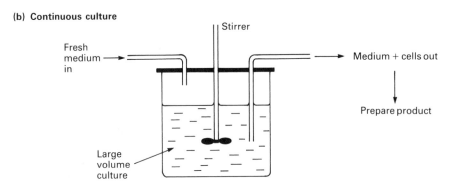

Figure 11.1 Two different systems for the growth of microorganisms. (a) Batch culture. (b) Continuous culture.

genes means that a gene for an important animal or plant protein can now be taken from its normal host, inserted into a cloning vector, and introduced into a bacterium (Figure 1.4). If the manipulations are performed correctly then the gene will be expressed and the protein synthesized by the bacterial cell. It may then be possible to obtain large amounts of the protein.

Of course, in practice obtaining **recombinant protein** is not as easy as it sounds. Special types of cloning vector are needed, and satisfactory yields of recombinant protein are often difficult to obtain. In this chapter we will look at cloning vectors for recombinant protein synthesis and examine some of the problems associated with their use.

11.1 GENES FROM HIGHER ORGANISMS ARE NOT NORMALLY EXPRESSED IN *E. coli*

A close look at the way in which genes are expressed will explain why a gene

Table 11.1 Some of the compounds produced on an industrial scale by culture of microorganisms

Compound	Microorganism
Antibiotics	
Penicillins	*Penicillium* spp.
Cephalosporins	*Cephalosporium* spp.
Gramicidins, polymixins	*Bacillus* spp.
Chloramphenicol, streptomycin	*Streptomyces* spp.
Enzymes	
Invertase	*Saccharomyces cerevisiae*
Proteases, amylases	*Bacillus* sp., *Aspergillus* spp.
Others	
Alcohol	*S. cerevisiae, S. carlsbergensis*
Glycerol	*S. cerevisiae*
Vinegar	*S. cerevisiae*, acetic acid bacteria
Dextran	*Leuconostoc* spp.
Butyric acid	Butyric acid bacteria
Acetone, butanol	*Clostridium* spp.
Citric acid	*Aspergillus niger*

from a higher organism is normally inactive when cloned into *E. coli*. In order to be expressed, a gene must be surrounded by a collection of signals that can be recognized by the bacterium. These signals, which are usually short sequences of nucleotides, advertise the presence of the gene and provide instructions for the transcriptional and translational apparatus of the cell. The three most important signals for *E. coli* are as follows (Figure 11.2).

1. The **promoter**, which marks the point at which transcription of the gene should start. In *E. coli* the promoter is recognized by the sigma factor, a component of the transcribing enzyme RNA polymerase.
2. The **terminator**, which marks the point at the end of the gene where transcription should stop. A termination signal is usually a short segment of DNA that can base-pair with itself to form a **stem–loop** structure.
3. The **ribosome binding site**, a short nucleotide sequence recognized by the ribosome as the point at which it should attach to the mRNA molecule. The initiation codon of the reading frame is always just a few nucleotides downstream of this site.

11.1.1 EXPRESSION SIGNALS ARE DIFFERENT IN DIFFERENT ORGANISMS

The reasons why genes from higher organisms will not work in bacterial cells became clear when DNA sequencing started to provide information about genes and their surrounding regions. Expression does not occur

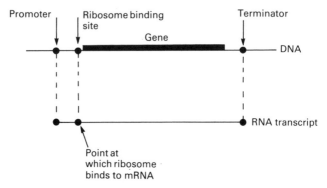

Figure 11.2 The three most important signals for gene expression in *E. coli*.

because the signals are different in different organisms. For example, *E. coli* genes are preceded by two separate sequences, TTGACA and TATAAT (or close variations), which combine to provide the start signal for transcription (Figure 11.3a). In animal cells the promoter also comprises two separate sequences, but these bear only a passing resemblance to the *E. coli* ones (Figure 11.3b); it is very unlikely that the *E. coli* RNA polymerase would be able to recognize the animal promoter sequences. Termination signals and ribosome binding sites are also different in bacteria and animals. Cloned animal genes are not expressed in *E. coli* quite simply because the enzymes of the bacterium do not recognize the regulatory sequences of the animal genes.

11.2 EXPRESSION VECTORS

Once the importance of these signals had been recognized, a way could be devised for expressing foreign (i.e. non-bacterial) genes in *E. coli*. If a foreign gene can be inserted into a vector in such a way that it is placed under control

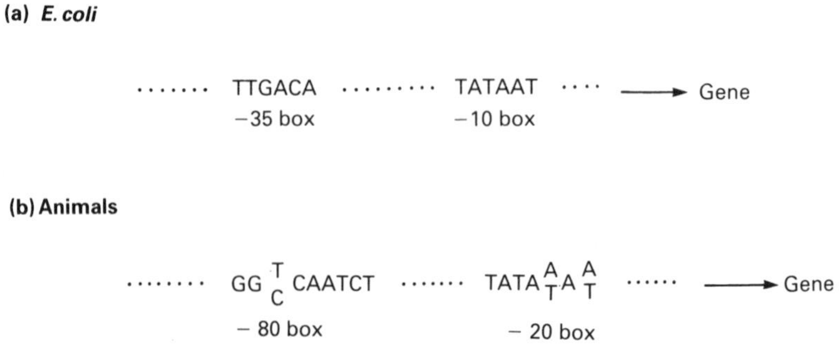

(a) *E. coli*

```
·······  TTGACA  ········  TATAAT  ····  ——→ Gene
          −35 box           −10 box
```

(b) Animals

```
                T
········  GG    CAATCT  ·······  TATA A A  ······  ——→ Gene
                C                      T  T
          − 80 box                − 20 box
```

Figure 11.3 Typical promoter sequences for *E. coli* and animal genes.

of *E. coli* expression signals, then that gene should be transcribed and translated (Figure 11.4). Cloning vehicles which provide these signals, and which can be used in the production of recombinant protein, are called expression vectors.

11.2.1 THE SIMPLEST EXPRESSION VECTORS CARRY JUST AN *E. coli* PROMOTER

The simplest expression vectors provide just one of the signals needed for gene expression, the promoter that directs initiation of transcription in *E. coli*. These vectors are designed so that the promoter sequence, taken

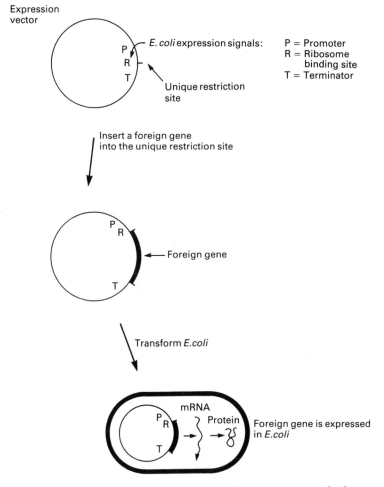

Figure 11.4 The use of an expression vector to achieve expression of a foreign gene in *E. coli*.

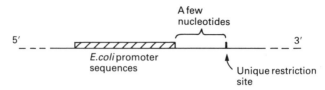

Figure 11.5 The relative positions on a simple expression vector of the promoter and the unique restriction site for insertion of a new gene.

from the upstream region of an *E. coli* gene, is positioned just in front of a unique restriction site into which new DNA can be inserted (Figure 11.5). A gene present on a fragment inserted into this restriction site will be placed under the control of the *E. coli* promoter and will therefore be transcribed in the bacterium.

11.2.2 THE PROMOTER MUST BE CHOSEN WITH CARE

The two short sequences shown in Figure 11.3a are **consensus sequences**, averages of all the *E. coli* promoter sequences that are known. Although most *E. coli* promoters do not differ much from these consensus sequences (e.g. TTGACT instead of TTGACA), a small variation may in fact have a major effect on the efficiency with which the promoter can direct transcription. **Strong promoters** are those that can sustain a high rate of transcription; strong promoters usually control genes whose translation products are required in large amounts by the cell (Figure 11.6a). In contrast, **weak promoters**, which are relatively inefficient, direct transcription of genes whose products are needed in only small amounts (Figure 11.6b). Clearly an expression vector should carry a strong promoter, so that the cloned gene is transcribed at the highest possible rate.

A second factor to be considered when constructing an expression vector is whether it will be possible to regulate the promoter in any way. Two major types of gene regulation are recognized in *E. coli* – **induction** and **repression**. An inducible gene is one whose transcription is switched on by addition of a chemical to the growth medium; often this chemical will be one of the substrates for the enzyme coded by the inducible gene (Figure 11.7a). In contrast, a repressible gene is switched off by addition of the regulatory chemical (Figure 11.7b).

Gene regulation is a complex process that only indirectly involves the promoter itself. However, many of the sequences important in induction and repression lie in the region immediately upstream of the gene and will be present, along with the promoter, on the piece of DNA carried by an expression vector. It may therefore be possible to extend the regulation to the expression vector itself, so that the chemical that induces or represses the gene normally controlled by the promoter is also able to regulate expression of the cloned gene.

(a) A strong promoter

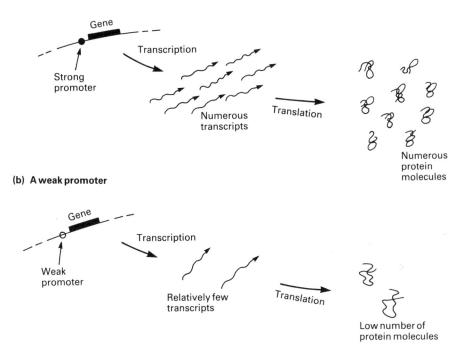

(b) A weak promoter

Figure 11.6 Strong and weak promoters.

This can be a distinct advantage in the production of recombinant protein. For example, if the recombinant protein has a harmful effect on the bacterium, then its synthesis must be controlled carefully to prevent accumulation of toxic levels; this can be achieved by judicious use of the regulatory chemical to control expression of the cloned gene. But even if the recombinant protein has no harmful effects on the host cell, regulation of the cloned gene is still desirable, as a continuously high level of transcription may affect the ability of the recombinant plasmid to replicate, leading to its eventual loss from the culture.

11.2.3. EXAMPLES OF PROMOTERS USED IN EXPRESSION VECTORS

Several *E. coli* promoters combine the desired features of strength and ease of regulation. Those most frequently used in expression vectors are as follows.

1. The *lac* **promoter** (Figure 11.8a), which is the sequence that controls transcription of the *lacZ* gene coding for β-galactosidase (and also the

(a) An inducible gene

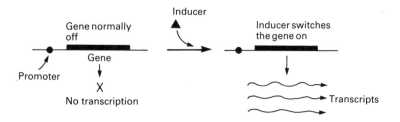

(b) A repressible gene

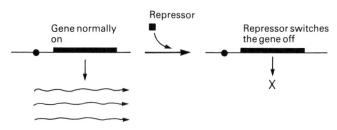

Figure 11.7 Examples of the two major types of gene regulation that occur in bacteria. (a) An inducible gene. (b) A repressible gene.

lacZ' gene fragment carried by the pUC and M13mp vectors – pp. 108 and 111). The *lac* promoter is induced by IPTG (p. 95), so addition of this chemical into the growth medium will switch on transcription of a gene inserted downstream of the *lac* promoter carried by an expression vector.

2. The **trp promoter** (Figure 11.8b), normally upstream of the cluster of genes coding for several of the enzymes involved in biosynthesis of the amino acid tryptophan. The *trp* promoter is repressed by tryptophan, but is more easily induced by 3-indolylacetic acid.

3. The **tac promoter** (Figure 11.8c), a hybrid between the *trp* and *lac* promoters that is in fact stronger than either, but still induced by IPTG.

4. The λP_L promoter (Figure 11.8d), one of the promoters responsible for transcription of the λ DNA molecule. λP_L is a very strong promoter that is recognized by the *E. coli* RNA polymerase, which is subverted by λ into transcribing the bacteriophage DNA. The promoter is repressed by the product of the λcI gene. Expression vectors that carry the λP_L promoter are used with a mutant *E. coli* host that synthesizes a temperature-sensitive form of the *cI* protein (p. 42). At a low temperature (less than 30°C) this mutant *cI* protein is able to repress the λP_L promoter; at higher temperatures the protein is inactivated resulting in transcription of the cloned gene.

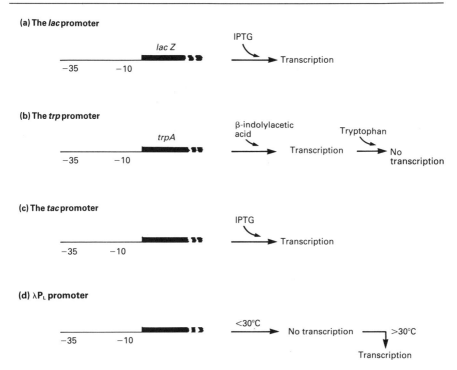

Figure 11.8 Four promoters often used in expression vectors. The *lac* and *trp* promoters are shown upstream for the genes that they normally control in *E. coli*.

11.2.4 USES OF TRANSCRIPTIONAL EXPRESSION VECTORS

At first glance it may seem pointless to construct a transcriptional expression vector that carries only an *E. coli* promoter, thereby ignoring the other important signals needed for expression of a foreign gene in a bacterium. In fact these vectors have numerous uses in molecular biology and biotechnology.

Their major role is in the study of *E. coli* genes. If an *E. coli* gene controlled by a weak promoter is inserted, along with its expression signals, into a normal cloning vector, then it will continue to be expressed at a relatively low level in the recombinant cells, even if present on a multicopy plasmid. This will be a disadvantage if the aim of the cloning experiment is to study the translation product of the cloned gene. However, if the gene is inserted into a transcriptional vector, and thereby put under the control of a stronger promoter, then it will be expressed at a high rate and its translation product will be synthesized in larger amounts. The gene's own terminator and ribosome binding site will of course provide the signals lacking in the expression vector.

A transcriptional expression vector may also be useful for the expression of some foreign genes. Transcription initiation is often the critical stage in gene expression and in some cases the other signals will not be so important. The terminator in particular may be dispensable as transcription will usually stop eventually, due to the RNA polymerase non-specifically detaching from the DNA (although the absence of a terminator can lead to some very undesirable effects). The *E. coli* ribosome binding site may also be unnecessary, as the foreign gene may possess or have attached to it a sequence that the *E. coli* ribosome can recognize and bind to. A low level of expression may therefore be obtained from a foreign gene inserted into a transcriptional vector.

However, much more efficient expression systems would be an advantage in biotechnology. More sophisticated vectors, ones that provide all the necessary expression signals, must therefore be considered.

11.3 CASSETTE VECTORS PROVIDE ALL THE EXPRESSION SIGNALS

In recent years a more advanced generation of expression vectors have been developed. These provide all the signals needed for gene expression (promoter, terminator and ribosome binding site) in the form of a **cassette**, so called because the foreign gene is inserted into a unique restriction site present in the middle of the expression signal cluster (Figure 11.9a).

A cassette vector is constructed by first inserting an entire *E. coli* gene, plus expression signals, into the vector. Manipulative techniques are then used to remove the reading frame of the cloned gene, leaving just the expression signals (Figure 11.9b). These manipulations must be designed and performed with great care if the signals are to be left intact. As with transcriptional expression vectors, the *E. coli* gene used in construction of a cassette is usually one with a strong, regulatable promoter.

The most satisfactory type of cassette vector is one from which the entire bacterial reading frame has been removed, leaving the expression signals and nothing else. A foreign gene inserted into such a cassette will be transcribed and the mRNA translated, with protein synthesis starting at the first methionine codon downstream of the ribosome binding site. The only stipulation is that this methionine codon must not be too far away from the ribosome binding site – if the two are separated by more than about 20 nucleotides then the efficiency of translation will be severely reduced.

In practice, cassette vectors of this type are difficult to develop, mainly because of the problems inherent in removing the entire *E. coli* reading frame while leaving the expression signals intact. Oligonucleotide synthesis is improving this situation, as promoter, terminator and ribosome binding signals can now be synthesized in the test-tube as short oligonucleotides, ligated together to form a cassette, and then inserted into a plasmid, producing a made-to-measure expression vector. However, this approach is very new. Cassette vectors of a second type, easier to construct because they retain some of the *E. coli* reading frame, are still of great importance in the production of recombinant protein.

(a) A cassette vector

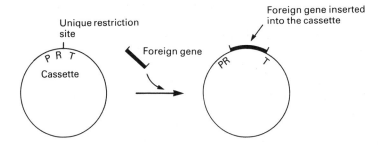

(b) Construction of cassette vector

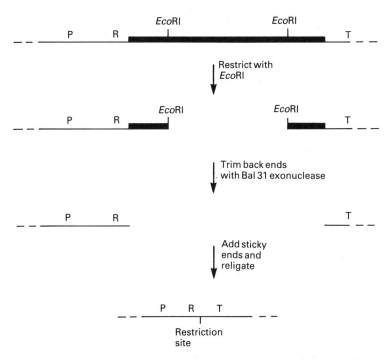

Figure 11.9 Cassette vectors. (a) A typical cassette vector and the way it is used. (b) One method for the construction of a cassette vector. Abbreviations: P = promoter, R = ribosome binding site, T = terminator.

11.3.1 CASSETTE VECTORS THAT SYNTHESIZE FUSION PROTEINS

With these vectors the unique restriction site used for insertion of new DNA lies within the reading frame of the original *E. coli* gene. Usually most, but not all, of this reading frame has been removed, and the unique site lies fairly near the start of the gene. Insertion of a new gene into this restriction site must be performed in such a way as to fuse the two reading frames (Figure 11.10), producing a hybrid gene, starting with the first few codons of the original *E. coli* reading frame, and progressing without a break into the codons of the foreign gene. The product of gene expression will therefore be a hybrid protein, consisting of the short peptide coded by the *E. coli* reading frame fused to the amino-terminus of the foreign gene.

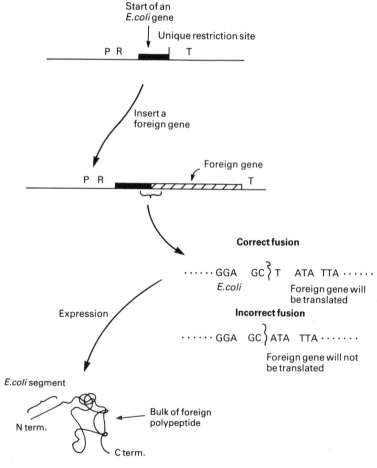

Figure 11.10 The construction of a hybrid gene and the synthesis of a fusion protein. Abbreviations: P = promoter, R = ribosome binding site, T = terminator.

(a) Premature termination of transcription

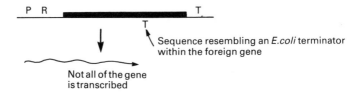

Sequence resembling an *E.coli* terminator
within the foreign gene

Not all of the gene
is transcribed

(b) *E.coli* cannot excise introns

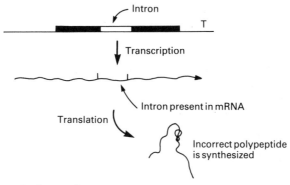

Intron

Transcription

Intron present in mRNA

Translation

Incorrect polypeptide
is synthesized

(c) Glycosylation—attachment of sugar molecules

Functional animal
protein with sugar
groups attached
to various amino acids

Non-glycosylated
recombinant protein

Figure 11.12 Three of the problems that may be encountered when foreign genes are expressed in *E. coli*. (a) Premature termination of transcription. (b) Introns are not excised in *E. coli*. (c) Proteins are not glycosylated in *E. coli*. Abbreviations: P = promoter, R = ribosome binding site, T = terminator.

(b) *Processing.* The proteins of most organisms are processed after translation, by the addition of chemical groups and by the modification of amino acids within the polypeptide. Unfortunately the proteins of bacteria and higher organisms are not processed in the same way. In particular, many animal proteins have sugar groups attached to them after translation (Figure 11.12c). These groups may be essential for the correct functioning of the protein, but are not added by E. coli to recombinant versions of the protein.

5. *At the level of the cell.* Expression of the cloned gene may have a harmful effect on the host cell, even if the translation product itself is not toxic. The energy and molecular drain caused by the rapid expression of a cloned gene may put a recombinant bacterium at a selective disadvantage, in terms of growth rate and cell stability, when compared with a non-recombinant cell. Bacteria whose plasmids have been altered or lost, so that they are no longer expressing the cloned gene, can easily outgrow the recombinants in a large-volume culture, resulting in eventual loss of product formation.

Whether or not recombinant protein can be produced in significant quantities in E. coli depends very much on the individual gene concerned. With some genes the system has worked reasonably well, with others the problems have been insurmountable.

11.5 PRODUCTION OF RECOMBINANT PROTEIN BY EUKARYOTIC CELLS

The problems associated with obtaining high yields of recombinant protein from genes cloned in E. coli have led to the development of expression systems for higher organisms. The argument is that a microbial eukaryote, such as a yeast or filamentous fungus, is more closely related to an animal or plant, so may be able to deal with recombinant protein synthesis more efficiently than E. coli. Yeasts and fungi can be grown just as easily as bacteria in continuous culture, and may express a cloned gene from a higher organism, and process the resulting protein, in a manner more akin to that occurring in the higher organism itself.

11.5.1 RECOMBINANT PROTEIN FROM YEAST AND FILAMENTOUS FUNGI

To a certain extent these hopes have been realized and microbial eukaryotes are now being used for the routine production of some animal proteins. Expression vectors are still required as it turns out that the promoters and other expression signals for animal genes do not in general work efficiently in these lower eukaryotes. The vectors themselves are based on those described in Chapter 7.

With *Saccharomyces cerevisiae*, the *GAL* promoter (Figure 11.13a) is frequently used for expression of foreign genes. This promoter is normally upstream of the gene coding for galactose epimerase, an enzyme involved in the metabolism of galactose. The *GAL* promoter is induced by galactose, providing a straightforward system for regulating expression of a cloned foreign gene. Amongst the filamentous fungi, *Aspergillus nidulans* has been

(a) The *GAL* promoter

(b) The glucoamylase promoter

(c) The cellobiohydrolase promoter

Figure 11.13 Three promoters frequently used in expression vectors for microbial eukaryotes.

the most popular choice so far, with the glucoamylase promoter (Figure 11.13b), induced by starch and repressed by xylose, frequently being used in expression vectors. *A. nidulans* may not however be the best fungus for recombinant protein production: the highest yield of a foreign protein so far achieved in a fungal host has been with an animal gene cloned in the cellulose-degrading fungus *Trichoderma reesei*, in an expression vector carrying the promoter from the cellobiohydrolase gene (Figure 11.13c).

Yeast and filamentous fungi seem to process foreign proteins in a manner similar to higher organisms, though it can be quite difficult to assess if chemical modifications such as glycosylation have been carried out correctly. With yeast, codon bias (p. 238) may be a problem, as those natural yeast genes that are expressed at the highest rate display a marked bias towards certain codons. For example, when the sequence of the highly expressed yeast gene coding for phosphoglycerol kinase is examined, it is found that of the six possible codons for leucine, TTG is used nine times out of ten and CTG and CTC are never used. A similar bias is seen with other highly expressed genes in yeast, suggesting that foreign genes, which rarely have such a codon bias, will not be expressed efficiently. Yeast also has the problem that the expressed protein is generally transported to the cell vacuole, rather than being secreted into the growth medium (Figure 11.14). In contrast, filamentous fungi are extremely efficient at protein secretion, and for this reason are rapidly superseding yeast as the organisms of choice for the biotechnological production of recombinant protein.

(a) Yeast transports recombinant protein to the cell vacuole

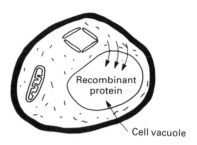

(b) Filamentous fungi secrete recombinant protein

Figure 11.14 One important difference between yeast and filamentous fungi with respect to production of recombinant protein.

11.5.2 USING HIGHER ORGANISMS FOR RECOMBINANT PROTEIN PRODUCTION

Biotechnologists are starting to realize that higher organisms may be the most efficient hosts for production of recombinant proteins. The rapid progress made in development of cloning vectors for plants and animals (Chapter 7) has been allied with similar advances in cell culture systems for higher organisms. Animal and plant cell culture methods have been around since the early 1960s but only recently have methods for large-scale continuous culture become available. In fact neither animal nor plant cells like to be completely suspended in culture media, preferring instead to be immobilized by attachment to some kind of solid support (Figure 11.15). The resulting cultures have much longer generation times than microbial cultures, limiting the yield of recombinant protein obtained from them, but this problem is more than offset by the fact that the recombinant protein will almost certainly be synthesized correctly.

Of course gene cloning may not be necessary in order to obtain an animal protein from an animal cell culture. Nevertheless, expression vectors and

(a) Animal cells in monolayer culture in a multiplate vessel

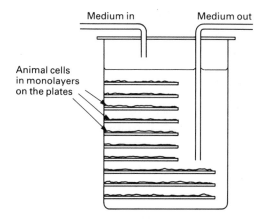

(b) Plant cells immobilized on foam particles

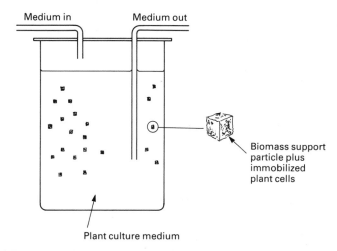

Figure 11.15 Immobilized cell culture techniques for (a) animal cells, and (b) plant cells.

cloned genes are still used to improve yields, by placing the gene under control of a promoter that is stronger than the one it is normally attached to. Two promoters that have been used in mammalian cells are the heat-shock promoter of the human *hsp-70* gene, which is induced at temperatures above 40°C, and the mouse metallothionein gene promoter, which is switched on by addition of zinc salts to the culture medium (Figure 11.16).

(a) The human heat-shock promoter

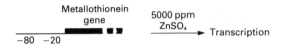

hsp-70

−80 −20

>40°C → Transcription → <40°C → No transcription

(b) The mouse metallothionein promoter

Metallothionein gene

−80 −20

5000 ppm ZnSO₄ → Transcription

Figure 11.16 Two promoters that have been used in expression vectors for mammalian cells.

Insect cells provide a potentially important alternative to mammalian cells for animal protein production. Insect cells do not behave in culture any differently than mammalian cells but have the great advantage that, thanks to a natural expression system, they can provide extremely high yields of recombinant protein. The expression system is based on the **baculoviruses**, a group of viruses that are common in insects but apparently do not infect vertebrates. The baculovirus genome includes a gene whose normal product accumulates in the insect cell as large crystalline bodies towards the end of the infection cycle (Figure 11.17). The product of this single gene frequently makes up over 50% of the total cell protein. Similar levels of protein production also occur when the normal gene is replaced by a foreign one. Insect cells appear to process mammalian proteins correctly, and indeed it is difficult at the moment to find any disadvantages with the baculovirus system. Further research will determine if this initial promise can be fulfilled.

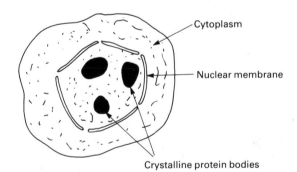

Cytoplasm

Nuclear membrane

Crystalline protein bodies

Figure 11.17 Crystalline protein bodies appear in the nuclei of insect cells infected with a baculovirus.

The final possibility to consider is one that makes use of an intact organism rather than a cell culture. The seeds of higher plants are very efficient at protein synthesis, as they accumulate large quantities of storage proteins and other compounds that the young seedling uses as a nutrient supply during the early stages of germination. Many crop plants have been bred specifically for the protein content of their seeds and the genes involved in seed development are quite well understood. What if a gene for a natural seed protein were to be replaced by a gene coding for some useful foreign protein? The answer is that under some circumstances the foreign protein will accumulate in the seeds, as has been demonstrated by the synthesis of pharmaceutical compounds called enkephalins in the seeds of engineered oilseed rape plants. Enkephalins are small proteins, only a few amino acids in length, and it still has to be established that larger foreign proteins can be synthesized efficiently in the special environment found within the developing seed. Further success in this area would open up an exciting new area of biotechnology.

FURTHER READING

Smith, J. E. (1985) *Biotechnology Principles*. Van Nostrand Reinhold, Wokingham, UK – includes sections on continuous culture and compounds produced on an industrial scale by microorganisms.

DeBoer, H. A. *et al.* (1983) The *tac* promoter: a functional hybrid derived from the *trp* and *lac* promoters. *Proceedings of the National Academy of Sciences, USA*, **80**, 21–5.

Remaut, E. *et al.* (1981) Plasmid vectors for high-efficiency expression controlled by the P_L promoter of coliphage. *Gene*, **15**, 81–93 – construction of an expression vector.

Grosjean, H. and Fiers, W. (1982) Preferential codon usage in prokaryotic genes: the optimal codon–anticodon interaction energy and the selective codon usage in efficiently expressed genes. *Gene*, **18**, 199–209 – the importance of codon bias in gene expression.

Saunders, G. *et al.* (1989) Heterologous gene expression in filamentous fungi. *Trends in Biotechnology*, **7**, 283–7.

Cameron, I. R. *et al.* (1989) Insect cell culture technology in baculovirus expression systems. *Trends in Biotechnology*, **7**, 66–70.

Vanderkerckhove, J. *et al.* (1989) Enkephalins produced in transgenic plants using modified 2S storage proteins. *Biotechnology*, **7**, 929–32.

gene cloning in medicine

12

The aim of this final chapter is to illustrate the impact that gene cloning is having in research and biotechnology, stimulating advances and break-throughs that could never be achieved without the ability to purify and manipulate individual genes. This is a topic that an entire book could be written about, but the endeavour would be a waste of time because the book would be out of date within a few weeks, so rapid is the current progress. Here we will limit ourselves to just one area in which the new biology is having a major influence and look at the way in which gene cloning is being applied in medicine. We will not even be able to take a comprehensive overview of medicine as a whole but instead will have to content ourselves with an examination of gene cloning in the production of pharmaceuticals, in the development of new types of vaccines, in the diagnosis of genetic diseases, and in the search for cures to the major diseases that afflict the human race.

12.1 PRODUCTION OF PHARMACEUTICAL COMPOUNDS

Many human disorders can be traced to the absence or malfunction of a protein normally synthesized in the body. An example is diabetes mellitus, which results when the hormone insulin is not synthesized in sufficient amounts by the β-cells of the Islets of Langerhans in the pancreas. Insulin controls the level of glucose in the blood, and an insulin deficiency manifests itself as a complex of symptoms which may lead to death if untreated. Fortunately many forms of diabetes can be alleviated by a continuing programme of insulin injections, thereby supplementing the limited amount of hormone synthesized by the pancreas. The insulin used in this treatment has traditionally been obtained from the pancreas of pigs and cows slaughtered for meat production. Although animal insulin is generally satisfactory, problems may arise in its use to treat human diabetes. One is that the slight differences between the animal and human proteins may lead to unpleasant side-effects in some patients. The second is that the

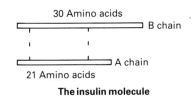

The insulin molecule

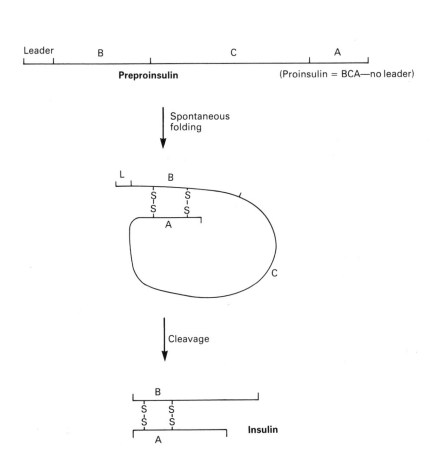

Figure 12.1 The structure of the insulin molecule and a summary of its synthesis by processing from preproinsulin.

purification procedures are difficult and potentially dangerous contaminants cannot always be completely removed.

A more satisfactory method would be to use human insulin for diabetes treatment. Obviously the protein cannot be prepared directly from humans, but could the gene be cloned and expressed in *E. coli*?

Insulin displays two features that facilitate its production by recombinant DNA techniques. The first is that the human protein is not modified after translation by the addition of sugar molecules (p. 239); recombinant insulin synthesized by a bacterium should therefore be active. The second advantage concerns the size of the molecule. Insulin is a relatively small protein, comprising two polypeptides, one of 21 amino acids (the A chain) and one of 30 amino acids (the B chain, Figure 12.1). In humans these are synthesized as a precursor called preproinsulin which contains the A and B segments linked by a third chain (C) and preceded by a leader sequence. The leader sequence is removed after translation and the C chain excised, leaving the A and B polypeptides linked to each other by two disulphide bridges.

Several strategies have been used to obtain recombinant insulin. One of these, involving synthesis of artificial genes for the A and B chains, illustrates a technique that has been of general utility in the production of recombinant protein in E. coli.

12.1.1 SYNTHESIS OF ARTIFICIAL INSULIN GENES

In the late 1970s the idea of making an artificial gene was extremely innovative. Oligonucleotide synthesis was in its infancy at that time, and the available methods for making artificial DNA molecules were much more cumbersome than the present-day automated techniques. Nevertheless genes coding for the A and B chains of insulin were synthesized as early as 1978.

The procedure used was to synthesize trinucleotides representing all the possible codons and then ligate these together in the order dictated by the amino acid sequences of the A and B chains. The artificial genes would not necessarily have the same nucleotide sequences as the real gene segments coding for the A and B chains, but would still specify the correct polypeptides. Two recombinant plasmids were then constructed, one carrying the artificial gene for the A chain and one the gene for the B chain. In each case the artificial gene was inserted into a lacZ' reading frame present on a pBR322-type vector (Figure 12.2a). The insulin genes were therefore under the control of the strong lac promoter (p. 231), and would be expressed as a fusion protein, consisting of the first few amino acids of β-galactosidase followed by the A or B polypeptide (Figure 12.2b). The genes were in fact designed so that the β-galactosidase and insulin segments would be separated by a methionine residue, the only ones present in either fusion protein. The insulin polypeptides could therefore be cleaved from the β-galactosidase segments by treatment with cyanogen bromide, which cuts proteins specifically at methionine residues. The purified A and B chains can then be attached to each other by disulphide bond formation in the test-tube.

This final step, involving disulphide bond formation, is actually rather inefficient. A subsequent improvement has been to synthesize not the individual A and B genes, but the entire proinsulin reading frame, specifying B chain – C chain – A chain (Figure 12.1). Although this is a more

(a) The artificial genes

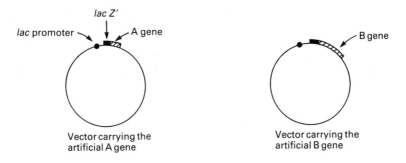

Vector carrying the
artificial A gene

Vector carrying the
artificial B gene

(b) Synthesis of insulin protein

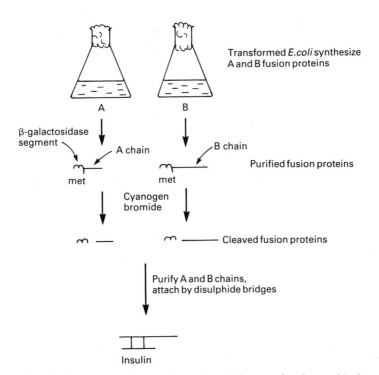

Figure 12.2 The synthesis of recombinant insulin from artificial A and B chain genes.

daunting proposition in terms of DNA synthesis, the prohormone has the big advantage of folding spontaneously into the correct, disulphide-bonded structure. The C chain segment can then be excised relatively easily by proteolytic cleavage.

12.1.2 SYNTHESIS OF OTHER ANIMAL PROTEINS IN *E. coli*

Insulin is perhaps the most notable success as far as recombinant protein production in *E. coli* is concerned. However, it is by no means the only animal protein to have been synthesized in *E. coli*. Three other examples, illustrating different approaches to the problems associated with expressing animal genes in a bacterium, are provided by somatostatin, somatotropin and interferon.

(a) Somatostatin – anti-growth hormone

This was the first human protein to be synthesized in *E. coli*. It is a small peptide hormone that acts in conjunction with a second protein, somatotropin, to regulate human growth. Somatostatin is important in the treatment of a variety of human growth disorders, including acromegaly, a condition characterized by uncontrolled bone growth.

The cloning of somatostatin was in fact a testing ground for the techniques later used in the synthesis of the artificial insulin genes. Being a very short protein, only 14 amino acids in length, somatostatin was ideally suited for artificial gene synthesis. The strategy used was the same as described for insulin, involving insertion of the artificial gene into a *lacZ'* vector (Figure 12.3), synthesis of a fusion protein, and cleavage with cyanogen bromide.

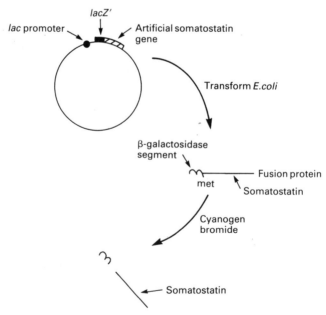

Figure 12.3 Production of recombinant somatostatin.

(b) Somatotropin – human growth hormone (HGH)

HGH is the complement to somatostatin and is also used in the treatment of human growth disorders, notably dwarfism.

Unfortunately HGH is not a small protein, being 191 amino acids long. This requires a gene of almost 600 nucleotides, a difficult prospect for the DNA synthesis capabilities of the late 1970s. In fact a combination of artificial gene synthesis and cDNA cloning was used to obtain an HGH-producing *E. coli* strain. HGH mRNA is present in the pituitary, the gland that produces the hormone in the human body. cDNA prepared from HGH mRNA has a unique site for the restriction endonuclease *Hae*III, which therefore cuts the molecule into two segments (Figure 12.4a). The longer

(a) Preparation of the HGH cDNA fragment

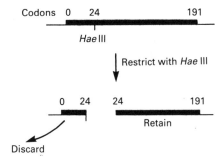

(b) HGH expression

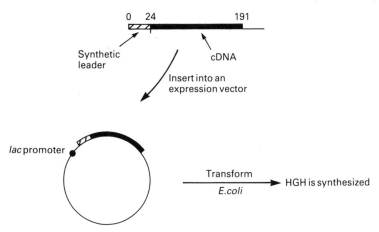

Figure 12.4 Production of recombinant human growth hormone (HGH).

segment, consisting of codons 24 to 191, was retained for use in construction of the recombinant plasmid. The smaller segment was replaced by an artificial DNA molecule that reproduced the start of the HGH gene, but provided the correct signals for translation in *E. coli* (Figure 12.4b). The modified gene was inserted into a transcriptional expression vector placing it under control of the *lac* promoter.

(c) Interferons

The interferons are a complex family of proteins that are synthesized by animal cells in response to attack by viruses. For many years it has been hoped that purified interferons may be useful as antiviral agents in the treatment of diseases as diverse as cancer and the common cold.

The major problem with the synthesis of interferons in *E. coli* has been that until quite recently these proteins had not been characterized in any great detail. Without the amino acid sequence, an artificial gene could not be

(a) Preparation of the cDNA library

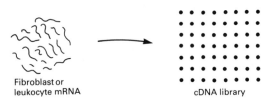

Fibroblast or
leukocyte mRNA

cDNA library

(b) Probing for interferon clones

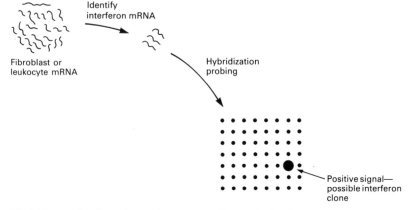

Fibroblast or
leukocyte mRNA

Identify
interferon mRNA

Hybridization
probing

Positive signal—
possible interferon
clone

Figure 12.5 The method used to isolate an interferon cDNA clone.

made by DNA synthesis. The gene therefore had to be obtained by cDNA cloning. Interferon is synthesized by cells such as human fibroblasts and leukocytes, so a cDNA library prepared with mRNA from these cells will include clones specific for interferon (Figure 12.5a). As may be expected, only a very small proportion of these clones will be the correct ones, so a screening procedure is needed for the interferon cDNAs. In practice, a hybridization probe was isolated by identifying mRNAs that direct synthesis of interferon in cell-free translation systems (Figure 12.5b).

More recently the amino acid sequences of some interferons have been determined and artificial genes constructed. This was not easy because the gene is over 1000 bp in length. First of all, 67 short oligonucleotides were synthesized, each one specifying a different part of the gene. These were joined together to produce 11 larger fragments, which were in turn ligated to give four gene segments. The final step was to ligate these four segments together to form a complete artificial gene which was inserted into a vector in which it was transcribed under the control of the *lac* promoter.

12.1.3 SYNTHESIS OF PHARMACEUTICALS FROM GENES CLONED IN EUKARYOTES

Although a number of important pharmaceutical compounds have been obtained from genes cloned in *E. coli*, the general problems associated with using bacteria to synthesize foreign proteins (p. 238) have led to these organisms now being almost completely superseded by eukaryotes as the chosen hosts for recombinant protein synthesis. A list of some but by no means all of the human proteins produced in eukaryotic cells is shown in Table 12.1. For most of these proteins there are still problems in obtaining

Table 12.1 Some of the human proteins that have been synthesized from genes cloned in eukaryotic cells

Protein	Host	Vector
Insulin	Mouse cells	Integrated plasmid
Somatostatin	Monkey cells	SV40
Growth hormone	Mouse cells	Papillomavirus
Parathyroid hormone	Rat cells	Retrovirus
Chorionic gonadotropin	Monkey cells	SV40
Interferons	Yeast	2 μm plasmid
	Silkworm larvae	Baculovirus
	Mouse cells	Papillomavirus
	Aspergillus nidulans	Integrated plasmid
Interleukins	Insect cells	Baculovirus
	Aspergillus nidulans	Integrated plasmid
Tissue plasminogen activator	*Aspergillus nidulans*	Integrated plasmid
Factor VIII	Hamster cells	Integrated plasmid

fully active products in sufficient quantities for the process to be commercially viable, but progress in this area is very rapid and a number of projects are now nearing the stage where marketable products will soon be available.

For many animal proteins the only possible way of obtaining them from cloned genes will be in a eukaryotic host. An example is provided by human Factor VIII, the last protein listed in Table 12.1. Factor VIII plays a central role in blood clotting, which is of course essential to prevent drastic injuries occurring through haemorrhaging. The commonest form of haemophilia in humans results from an inability to synthesize Factor VIII, leading to a breakdown in the blood clotting pathway and the well-known symptoms associated with the disease. At present the only way to treat haemophilia is by injection of purified Factor VIII protein, obtained from human blood provided by donors. Purification of Factor VIII is a complex procedure and the treatment is very expensive, with estimates of around $10 000 per annum for each patient. More critically, the purification is beset with difficulties, in particular in removing virus particles that may be present in the blood. Hepatitis and AIDS can and have been passed on to haemophiliacs via Factor VIII injections. Recombinant Factor VIII, free from contamination problems, would be a significant advance for gene cloning technology.

The problem with Factor VIII is that the gene is very large, over 186 kb in length, and is split into 26 exons and 25 introns (Figure 12.6). A complete cDNA for the gene would be 9 kb, which would press the current cDNA synthesis techniques to their limits. Furthermore, the Factor VIII protein contains 17 disulphide bonds and is active only after extensive glycosylation. Correct processing will almost certainly not occur in a non-mammalian host. Despite these various difficulties, human Factor VIII has been obtained from a gene cloned in recombinant hamster cells. Initially the yields of protein were very small but research into the processing events involved in synthesis of active Factor VIII identified important cofactors that the hamster cells were lacking. Once these were supplied the levels of Factor VIII increased dramatically, enough to make production on a large scale a commercially viable proposition. Two biotechnology companies in the United States are currently carrying out clinical trials with recombinant Factor VIII and if these are successful we can expect to see this latest triumph of gene cloning on the market in the near future.

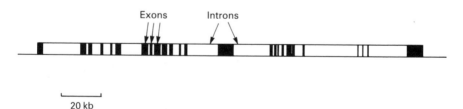

Figure 12.6 The structure of the human Factor VIII gene. Exons are shown as closed boxes and introns as open boxes.

12.2 PRODUCTION OF RECOMBINANT VACCINES

Over the last century, many infectious diseases, previously capable of devastating populations, have been brought under control by vaccination programmes. These make use of an antigenic preparation, called a vaccine, that after injection into the bloodstream stimulates the immune system to synthesize antibodies that will protect the body against subsequent infection.

The antigenic material present in a vaccine is normally an inactivated form of the infectious agent. Vaccines for viral diseases are virus particles that have been attenuated, so that they are no longer infective, by heating or a similar treatment. In the past, two problems have hindered the preparation of attenuated vaccines.

1. The inactivation process must be 100% efficient, as the presence in a vaccine of just one live virus particle could result in infection. This has been a problem with vaccines for the cattle disease, foot-and-mouth.
2. The large amounts of virus particles needed for vaccine production are usually obtained from tissue cultures. Unfortunately some viruses, notably hepatitis B virus, do not grow in tissue culture.

12.2.1 A NEW GENERATION OF RECOMBINANT VACCINES

Recombinant DNA technology offers a means of obtaining a new type of vaccine, one that will not suffer from the problems associated with attenuated viruses. The use of gene cloning in this field centres on the discovery that virus-specific antibodies may be synthesized in response not only to the whole viral particle, but also to isolated components of the virus. This is particularly true of purified preparations of the proteins present in the virus coat (Figure 12.7). Two methods are being developed for the production of recombinant vaccines.

(a) Cloning in expression vectors

If the genes coding for the antigenic proteins of a particular virus could be identified and inserted into an expression vector, then the methods described above for synthesis of animal proteins could be employed in the production of recombinant vaccines. This strategy has been successful with the viruses for foot-and-mouth disease and hepatitis B.

1. **Foot-and-mouth disease.** Genes for two foot-and-mouth disease virus proteins, VP1 and VP3, have been expressed in *E. coli*. In both cases several million protein molecules per cell were obtained, but unfortunately only VP3 has proved particularly useful as a vaccine.
2. **Hepatitis B.** This virus consists of two proteins, the core protein that is associated with the virus genome, and the coat protein or major surface antigen. The core protein has been successfully produced in large

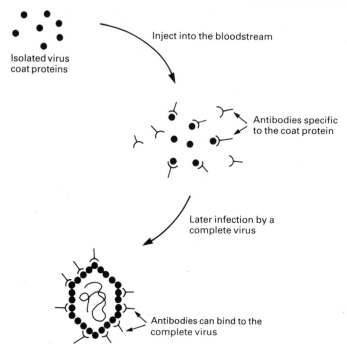

Figure 12.7 The principle behind the use of a preparation of isolated virus coat proteins as a vaccine.

quantities in recombinant *E. coli*; unfortunately this protein is not useful as a vaccine, although it has a valuable role in diagnosis of the disease. The major surface antigen, which would be a useful vaccine, was not synthesized in particularly large amounts in *E. coli*. However, greater success was achieved with the major surface antigen gene cloned in yeast, using a vector based on the 2 μm plasmid (p. 126). The major surface antigen protein was obtained in reasonably high quantities (25 μg per litre of culture) and when injected into monkeys provided protection against hepatitis B. The yeast protein is now being marketed for use as a human vaccine, and the possibility of using hamster cells to produce an even better vaccine is being explored.

(b) Recombinant vaccinia viruses

The use of live vaccinia virus as a vaccine for smallpox dates back to 1796, when Jenner first realized that this virus, harmless to man, could stimulate immunity against the much more dangerous smallpox virus. The term 'vaccine' comes from vaccinia; its use resulted in the 'official' extinction of smallpox in 1980.

A much more recent idea is that recombinant vaccinia viruses could be used as live vaccines against other diseases. If a gene coding for a virus coat protein, for example the hepatitis B major surface antigen, is ligated into the vaccinia genome, under control of a vaccinia promoter, then the gene will be expressed (Figure 12.8). After injection into the bloodstream, replication of the recombinant virus will result not only in new vaccinia particles, but also in significant quantities of the major surface antigen. Immunity against both smallpox and heptatitis B will result.

This remarkable technique has considerable potential. Already recombinant vaccinia viruses expressing a number of foreign genes have been constructed (Table 12.2) and shown to confer immunity against the relevant

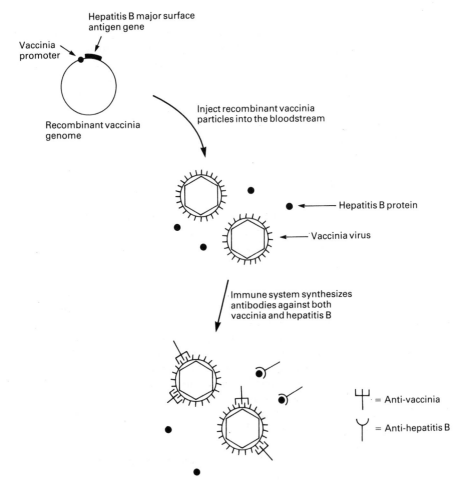

Figure 12.8 The rationale behind the potential use of a recombinant vaccinia virus.

Table 12.2 Foreign genes expressed by recombinant vaccinia viruses

Gene
Plasmodium falciparum surface antigen
Influenza virus coat proteins
Rabies virus G protein
Hepatitis B major surface antigen
Herpes simplex glycoproteins
HIV envelope proteins
Vesicular stomatitis coat proteins
Sindbis virus proteins

diseases in experimental animals. There is also the possibility of broad-spectrum vaccines with the demonstration that a single recombinant vaccinia, expressing the genes for influenza virus haemagglutinin, hepatitis B major surface antigen and herpes simplex virus glycoprotein, confers immunity against each disease in monkeys. This raises the possibility of a human vaccine that would protect against a variety of different diseases with just one inoculation. The important question that must now be answered about recombinant vaccinia viruses is whether we know enough about viral biology to be sure that their release into the ecosphere via vaccination programmes can be allowed.

12.3 SCREENING FOR GENETIC DISEASES

The progress that has been made in the production of recombinant pharmaceuticals and vaccines is mirrored by equally rapid advances in other applications of gene cloning. This is particularly true with respect to the use of recombinant DNA techniques in screening programmes for genetic diseases.

A genetic disease is one that is caused by a specific defect in an individual gene. Examples of some of the most important genetic diseases in the UK are given in Table 12.3. As human beings possess two copies of each chromosome (except for the X and Y chromosomes in males) a person can be a 'carrier' of a defective gene without ever being in danger of developing a genetic disease. This is because most defects of this kind are recessive, and the defective gene will be masked by the second, correct version of the gene (Figure 12.9). An exception is when the defective gene lies on the X chromosome, as then only females can be carriers: males only have one X chromosome so a male carrying a defective gene on the X chromosome will display the disease.

As genetic diseases are inherited it is generally possible to know if a couple are at risk of passing a disease to their children. In such cases the early diagnosis of the status of an unborn child will allow the couple to make a

Table 12.3 Some of the commonest genetic diseases in the UK

Disease	Frequency (births)
Cystic fibrosis	1 in 2000
Huntington's chorea	1 in 2000
Muscular dystrophy	1 in 5000
Phenylketonuria	1 in 10 000
Sickle cell anaemia	1 in 10 000
β-Thalassaemia	1 in 20 000
Retinoblastoma	1 in 20 000
Cystinuria	1 in 20 000
Tay–Saxs disease	1 in 200 000

choice, should they wish, about terminating the pregnancy. Prenatal diagnosis of some genetic diseases has been possible for several years through, for example, direct examination of chromosomes for characteristic cytogenetic abnormalities, or enzyme assays, which may indicate an inherited metabolic defect. However, these approaches are limited in scope, and with each it is necessary to culture fetal cells from the amniotic fluid, meaning that a reliable diagnosis cannot be obtained until relatively late in the pregnancy.

12.3.1 GENETIC SCREENING BY RECOMBINANT DNA METHODS

The ability to clone genes has led to a new approach to the screening of genetic diseases. DNA is obtained from fetal cells and examined directly for the presence of the correct or the defective copy of the relevant gene. Sufficient amounts of DNA for analysis can be obtained from fetuses as young as ten weeks. In fact very little DNA is needed as the region containing the gene can often be amplified by the polymerase chain reaction

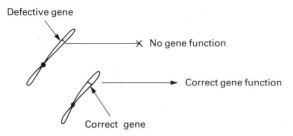

Figure 12.9 If a defective gene is recessive then its effect will be masked by the correct copy of the gene present on the other member of the chromosome pair. A recessive genetic disease is therefore not manifested in a heterozygous individual, but this individual will 'carry' the disease and could pass it on to his or her children.

(p. 197). Once DNA has been obtained there are three different strategies that can be used to gain information about the gene of interest.

(a) RFLP analysis for a specific gene defect

Some mutations will result in the alteration or deletion of a restriction site, leading to an RFLP (p. 195) that can be detected by the method shown in Figure 9.13. The presence of the RFLP will be a direct indication that the gene is defective and that the fetus will suffer from the resulting genetic disease. Unfortunately only a few genetic diseases can be diagnosed in this way, not so much because genetic defects that result in an RFLP are rare, but because as yet we do not know enough about the structures of the relevant genes to know what RFLPs to look for. Despite this, the method has proved useful for diagnosing various types of thalassaemia, which result from defects in the globin genes, and also in the diagnosis of some types of haemophilia.

(b) RFLP linkage analysis

An alternative and more generally useful method for genetic screening is **RFLP linkage analysis**. This does not depend on the existence of an RFLP that is a direct consequence of the genetic defect, but requires merely that a recognizable RFLP is present somewhere in the vicinity of the defective gene. If this RFLP is close enough then it will be inherited along with the defective gene, as it is very unlikely that a recombination event during meiosis will separate the RFLP from the gene. The presence of the RFLP will therefore be indicative of the presence of the defective gene.

For some genetic diseases we know of an RFLP that is very closely linked to the relevant gene. With these the chances that the RFLP and defective gene will ever be separated by recombination is vanishingly small, and the presence of the RFLP can be looked on as a confident diagnosis for the presence of the defective gene. One or two RFLPs of this nature have been identified by chance, but most are known because the relevant genes have been cloned and sequenced. For other genetic diseases we do not as yet have RFLPs that are really closely linked to the defective gene. With these diseases it is necessary to examine small high-risk populations to try and identify a suitable linked RFLP, possibly one that is applicable to just a single family group. This is a more difficult approach as often there are not enough living members of the group from which DNA samples can be obtained. Furthermore, it will be very difficult to provide such a screening service for every single risk group.

(c) Oligonucleotide probing for specific gene defects

In the long term, oligonucleotide probing is likely to become the main method used in genetic screening. This is because oligonucleotide probing can diagnose any genetic defect, so long as the precise nature of that defect is known. Consider the various possibilities. If the defect is due to a deletion then it will be detectable because there will be no hybridization to an oligonucleotide probe specific for the deleted region (Figure 12.10a). If the defect is due to a point mutation then it will be detected by an oligonucleotide probe designed in accordance with the correct nucleotide sequence, so long as the probe is hybridized to the target DNA under such conditions that a single base-pair mismatch will result in instability and no hybridization signal (Figure 12.10b). Finally, if the defect is due to an insertion then it will be detectable with a probe specific for a junction between the gene and the insertion, again with highly stringent conditions used so that the mismatches caused by the insertion will result in an unstable hybrid (Figure 12.10c).

The development of oligonucleotide probes for individual genetic diseases depends on a knowledge of the precise defect responsible for the disease. To obtain this knowledge the relevant gene must be cloned and sequenced. As well as resulting in an oligonucleotide probe the sequence of the gene may indicate the biochemical nature of the genetic defect, possibly pointing the way to a treatment and maybe even a cure for the disease itself. During the last few years molecular biologists have made progress in identifying and characterizing the genes for a number of genetic diseases. To

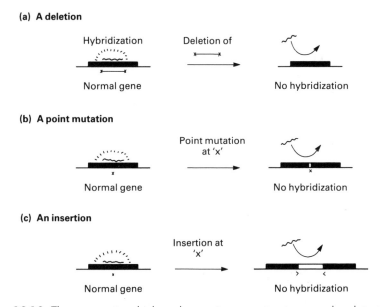

Figure 12.10 Three ways in which a change in gene structure can be detected by hybridization analysis with a suitable oligonucleotide probe. (a) A deletion can be detected with an oligonucleotide specific for the deleted region. (b) A point mutation can be detected with an oligonucleotide specific for the unmutated sequence. (c) An insertion can be detected with an oligonucleotide specific to the sequence in which the insertion occurs.

see what this work involves, and how valuable the information obtained can be, we will briefly examine the research programme that has led to identification of the gene which, when mutated, causes cystic fibrosis.

12.4 IDENTIFICATION OF THE CYSTIC FIBROSIS GENE

Cystic fibrosis is one of the commonest genetic diseases among caucasians: about one child in every 2000 born in the UK has cystic fibrosis (Table 12.3), one person in 20 is a carrier of the defective gene, and some 5000 people die throughout the world every year from the disease. Individuals with cystic fibrosis produce large quantities of mucus in their lungs, leading to respiratory problems and a high susceptibility to lung infection. There is no cure and a child with the disease is unlikely to reach adulthood.

12.4.1 THE CELLULAR AND BIOCHEMICAL BASIS TO CYSTIC FIBROSIS

Cystic fibrosis has of course been studied for many years and only recently has molecular biology been applied to the problem. Cell biologists and biochemists have known for some time that the key to cystic fibrosis lies with the epithelial cells of afflicted individuals. These cells line the surfaces of tissues and organs such as the lungs and intestinal tract, and also form the sweat and salivary glands. In a cystic fibrosis patient the epithelial cells display a number of abnormalities, and in particular are unable to transport ions in the correct way. This ties in well with the major symptom of the disease, the accumulation of mucus in the bronchial tubes. Unfortunately, ion trarsport systems are difficult to study. If the basic system is not understood then how can the defect responsible for cystic fibrosis be determined?

12.4.2 LOCATING THE CYSTIC FIBROSIS GENE

In the absence of information concerning the nature of the gene product, it proved impossible to make a hybridization probe for the cystic fibrosis gene. This meant that the gene could not be identified in a human genomic library. The only alternative was to locate the gene by chromosome walking (p. 186), which could not be attempted until a second gene, closely linked to the cystic fibrosis gene, had been identified.

By analysing the inheritance patterns of the cystic fibrosis and other genes in affected families, a marker closely linked to the cystic fibrosis gene was eventually discovered. This marker is called *D7S15*, and is not in fact a gene, but a large segment of cloned DNA that exists in different forms, akin to an RFLP. The different polymorphic forms of *D7S15* can be distinguished from one another, and their inheritance patterns can be followed in the same way as for an ordinary gene or RFLP. One form of *D7S15* was found to be almost always inherited with the defective cystic fibrosis gene. A linkage had been established and it was now known that the cystic fibrosis gene is located near to *D7S15*, on the long arm of human chromosome 7.

Unfortunately, the cystic fibrosis gene is not so closely linked to *D7S15* that the latter can be used as a starting point for a chromosome walk. However, once the position of the cystic fibrosis gene was known it was much easier to identify other, more closely linked markers. Two were found that flanked the cystic fibrosis gene, called *MET* and *D7S8*, and two further ones, *D7S122* and *D7S340* were shown to be located in the same segment, extremely close to the cystic fibrosis gene (Figure 12.11). Now the chromosome walk could begin.

Because several markers had been identified in the relevant region the chromosome walk did not have to be in just one direction, but instead could start at any one of the cloned marker DNA positions, and progress in either direction. Gradually an ordered set of clones was built up in the region known to contain the cystic fibrosis gene. One or more of these clones must contain the gene itself, but how could the gene be recognized without a hybridization probe?

(a) Identifying the cystic fibrosis gene

Several strategies were available to identify the cystic fibrosis gene in the cloned region. The most useful approach was to use individual clones as hybridization probes to Southern transfers of DNA from other mammals. A clone that contains an important human gene will almost certainly hybridize to an equivalent gene in other mammalian species. Four sets of clones, representing four areas contained within the region under study, were found to cross-hybridize with DNA from cows, mice and hamsters. Each of these areas therefore contained a gene: which was the cystic fibrosis gene?

Biochemical studies of cystic fibrosis had suggested that the gene responsible for the disease is expressed only in certain tissues. If the expression pattern of one of the cloned genes matched that predicted for the cystic fibrosis gene, then the identification would be almost complete. The expression pattern of a gene can be determined by hybridizing the gene to

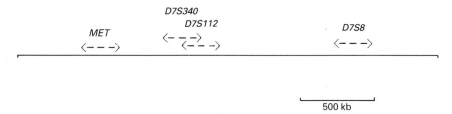

Figure 12.11 Genetic markers in the region of human chromosome 7 in which the cystic fibrosis gene is located. The cystic fibrosis gene was known to be close to *D7S340* and *D7S112* but its exact position was still a mystery.

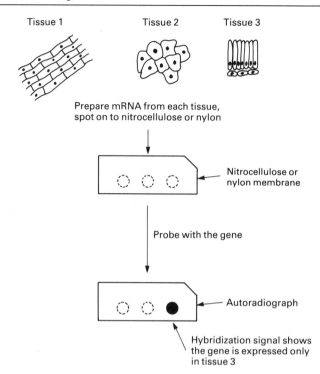

Figure 12.12 Dot blot hybridization analysis to work out the expression pattern of a cloned gene in different tissues.

mRNA or cDNA preparations from different tissues (Figure 12.12). A hybridization signal indicates the presence of transcripts in the tissue, showing that gene expression is occurring. Just one of the four genes identified in the cloned region had an expression pattern matching that expected for the cystic fibrosis gene. Was this the right gene?

Next the entire nucleotide sequence of the gene was determined. This proved to be a big task as the gene is made up of 24 exons spread over some 240 kb of DNA, making it one of the largest genes ever discovered (Figure 12.13). Eventually the work was completed and the amino acid sequence of the gene product could be predicted. Suddenly it was clear that the gene was the right one: the amino acid sequence possessed all the characteristic features of an ATP-dependent membrane transport protein.

12.4.3 THE GENETIC DEFECT RESPONSIBLE FOR CYSTIC FIBROSIS

To work out the exact nature of the defect responsible for cystic fibrosis it was necessary to compare the structures of the genes from afflicted and healthy individuals. This could have been done simply by sequencing the

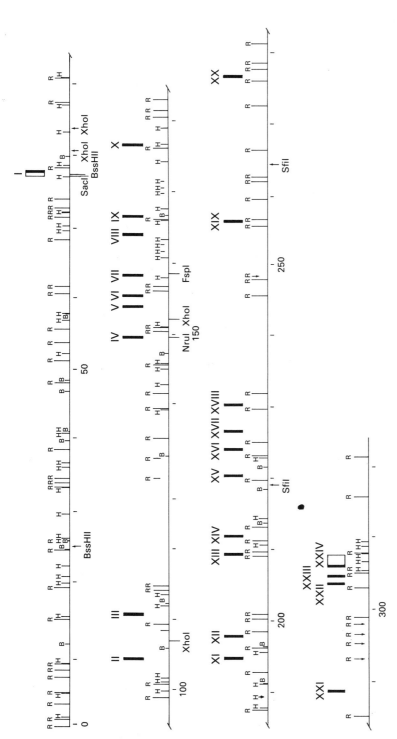

Figure 12.13 The structure of cystic fibrosis gene. The diagram shows 325 kb of human chromosome 7, with the positions of the 24 exons of the cystic fibrosis gene marked by solid vertical lines numbered with roman numerals. The positions of the leader and trailer regions are shown as open boxes attached to the first and last exons. The restriction map is also shown. Abbreviations: B, *Bam*HI; R, *Eco*RI; H, *Hind*III. Redrawn from J. M. Rommens *et al.* (1989) *Science*, **245**, 1059–65.

entire genes from two people, but a short cut was taken. The cloning work identified a number of RFLPs lying at various positions within the cystic fibrosis gene itself. These were used to carry out a very fine-scale RFLP linkage analysis, using DNA samples from the members of families with histories of cystic fibrosis, to pinpoint the region of the gene in which the mutation lies. Once the important region had been located, DNA sequencing was able to identify the mutation itself. Seventy per cent of all cystic fibrosis cases were found to be due to the deletion of three nucleotides, comprising a single phenylalanine codon, located within one of the ATP-binding domains of the protein (Figure 12.14).

Sometimes molecular biology projects fail to live up to expectations and the information provided is of limited use in defining the function of the gene being studied. Frequently a cloning and sequencing project leaves us no closer to understanding the biochemistry of the gene product and tells us only how the gene itself is constructed. This has not been the case with cystic fibrosis. The demonstration that the protein coded by the cystic fibrosis gene is a membrane transport protein has refocused the efforts of biochemists investigating the nature of the malfunction in cystic fibrosis cells. The fact that the amino acid that is deleted in the defective protein lies in a conserved, though not critical, region of the protein suggests that the defective protein is not completely inactive, but instead has a modified or reduced function. This is encouraging as a malfunction is generally easier to correct by biochemical means than is a total loss of activity. Most importantly, proteins related to the cystic fibrosis gene product are known in organisms other than

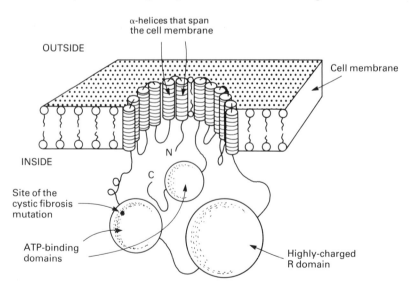

Figure 12.14 The proposed structure of the protein coded by the cystic fibrosis gene, showing the position of the mutation that occurs in 70% of cystic fibrosis patients. Redrawn from J. R. Riordan *et al.* (1989) *Science*, **245**, 1066–73.

man, and it is now clear that information gained with these other proteins will be of direct relevance to human cystic fibrosis. There is still a long way to go before a treatment or cure for cystic fibrosis is found, but gene cloning has pointed the way and can lead us further along the route.

FURTHER READING

Goeddel, D. V. *et al.* (1979) Expression in *Escherichia coli* of chemically synthesized genes for human insulin. *Proceedings of the National Academy of Sciences, USA,* **76**, 106–10.

Itakura, K. *et al.* (1977) Expression in *Escherichia coli* of a chemically synthesized gene for the hormone somatostatin. *Science,* **198**, 1056–63.

Goeddel, D. *et al.* (1979) Direct expression in *Escherichia coli* of a DNA sequence coding for human growth hormone. *Nature,* **281**, 544–8.

Gray, P. W. *et al.* (1982) Expression of human immune interferon cDNA in *E. coli* and monkey cells. *Nature,* **295**, 503–8.

Stahl, S. *et al.* (1982) Hepatitis B virus core antigen: synthesis in *Escherichia coli* and application in diagnosis. *Proceedings of the National Academy of Sciences, USA,* **79**, 1606–10.

Valuenzuela, P. *et al.* (1985) Synthesis and assembly in yeast of hepatitis B surface antigen particles containing the polyalbumin receptor. *Biotechnology,* **3**, 317–20.

Patzer, E. J. *et al.* (1986) Cell culture derived recombinant HBsAG is highly immunogenic and protects chimpanzees from infection with hepatitis B virus. *Biotechnology,* **4**, 630–6 – production in mammalian cell culture.

Tartaglia, J. and Paoletti, E. (1988) Recombinant vaccinia virus vaccines. *Trends in Biotechnology,* **6**, 43–6.

Gusella, J. F. *et al.* (1983) A polymorphic DNA marker genetically linked to Huntingdon's disease. *Nature,* **306**, 234–8 – an example of the use of RFLPs in diagnosis of genetic diseases.

Rommens, J. M. *et al.* (1989) Identification of the cystic fibrosis gene: chromosome walking and jumping. *Science,* **245**, 1059–65.

Further reading

The reader who has understood all the preceding chapters will now be in a position to tackle any one of several more advanced texts on gene cloning. The most useful is *From Genes to Clones: Introduction to Gene Technology* (1987) E.-L. Winnacker, VCH, Weinheim, FRG. This book can be recommended unreservedly to students seeking a more detailed treatment of gene cloning.

For readers who intend to become practitioners of gene cloning there are a number of cloning manuals that provide detailed protocols for the relevant techniques. The best ones are *Guide to Molecular Cloning Techniques* (1989) S. L. Berger and A. R. Kimmel, Academic Press, London, *Molecular Cloning*, 2nd edn (1989) by J. Sambrook *et al.*, Cold Spring Harbor Press, Cold Spring Harbor, New York, and *A Practical Guide to Molecular Cloning*, 2nd edn (1988) B. Perbal, Wiley Interscience, New York.

There are also one or two books that attempt to explain the applications of gene cloning in research and biotechnology. The most successful of these, giving a readable and accurate review of the subject as whole, is *Microbial Techniques Applied to Biotechnology* (1987) V. A. Saunders and J. R. Saunders, Croom Helm, London.

A less academic but nonetheless informative book is *Man Made Life* (1982) J. Cherfas, Basil Blackwell, Oxford. This describes the historical development of gene cloning, providing fascinating insights into the first cloning experiments and the early successes with recombinant protein. This book is particularly useful for its treatment of the moratorium placed on recombinant DNA experiments in the 1970s as a result of the Asilomar Conference.

Gene cloning techniques and biotechnology are advancing at a great pace and scientific journals offer the only means of keeping in touch with the developments. Both *Nature* and *New Scientist* regularly include articles and news on the latest applications of recombinant DNA technology. Two journals devoted to the industrial aspects of gene cloning, *Biotechnology* and *Trends in Biotechnology*, are also recommended.

Glossary

2 μm circle   A plasmid found in the yeast *Saccharomyces cerevisiae* and used as the basis for a series of cloning vectors.

Adaptor   A synthetic, double-stranded oligonucleotide used to attach sticky ends to a blunt-ended molecule.

Agrobacterium tumefaciens   The soil bacterium which, when containing the Ti plasmid, is able to form crown galls on a number of dicotyledonous plant species.

Autoradiography   A method of detecting radioactively labelled molecules through exposure of an X-ray sensitive photographic film.

Auxotroph   A mutant microorganism that will grow only if supplied with a nutrient not required by the wild-type.

Avidin   A protein that has a high affinity for biotin and is used in the detection system for biotinylated probes.

Bacteriophage or **Phage**   A virus whose host is a bacterium. Bacteriophage DNA molecules are often used as cloning vectors.

Baculovirus   A virus that has been used as a cloning vector for the production of recombinant protein in insect cells.

Batch culture   Growth of bacteria in a fixed volume of liquid medium in a closed vessel, with no additions or removals made during the period of incubation.

Biolistics   A means of introducing DNA into cells that involves bombardment with high-velocity microprojectiles coated with DNA.

Biological containment   One of the precautionary measures taken to prevent the replication of recombinant DNA molecules in microorganisms in the natural environment. Biological containment involves the use of vectors and host organisms that have been modified so that they will not survive outside of the laboratory.

Biotechnology   The use of living organisms, often but not always microbes, in industrial processes.

Biotin   A molecule that can be incorporated into dUTP and used as a non-radioactive label for a DNA probe.

Blunt end or **Flush end**   An end of a DNA molecule at which both strands terminate at the same nucleotide position with no single-stranded extension.

Bovine papillomavirus (BPV)   A group of mammalian viruses, derivatives of which have been used as cloning vectors.

Broth culture   Growth of microorganisms in a liquid medium.

Buoyant density   The density possessed by a molecule or particle when suspended in an aqueous salt or sugar solution.

Capsid The protein coat that encloses the DNA or RNA molecule of a bacterio-phage or virus.

Cassette A DNA sequence consisting of promoter – ribosome binding site – unique restriction site – terminator carried by certain types of expression vector. A foreign gene inserted into the unique restriction site will be placed under control of the expression signals.

Cauliflower mosaic virus (CaMV) The best studied of the caulimoviruses, used as a cloning vector for some species of higher plant.

Caulimoviruses One of the two groups of DNA viruses to infect plants, the members of which have potential as cloning vectors for some species of higher plant.

Cell extract A preparation consisting of a large number of broken cells and their released contents.

Cell-free translation system A cell extract containing all the components required for protein synthesis (i.e. ribosomal subunits, tRNAs, amino acids, enzymes and cofactors) and able to translate added mRNA molecules.

Chimaera A recombinant DNA molecule made up of DNA fragments from more than one organism, named after the mythological beast.

Chromosome A self-replicating nucleic acid molecule carrying a number of genes.

Chromosome walking A technique used to identify a series of overlapping restriction fragments, often to determine the relative positions of genes on large DNA molecules.

Cleared lysate A cell extract that has been centrifuged to remove cell debris, subcellular particles and possibly chromosomal DNA.

Clone A population of identical cells, generally those containing identical recombinant DNA molecules.

Compatibility Refers to the ability of two different types of plasmid to coexist in the same cell.

Competent Refers to a culture of bacteria that have been treated to enhance their ability to take up DNA molecules.

Complementary Refers to two polynucleotides that can base-pair to form a double-stranded molecule.

Complementary DNA (cDNA) cloning A cloning technique involving conversion of purified mRNA to DNA before insertion into a vector.

Conformation The spatial organization of a molecule. Linear and circular are two possible conformations of a polynucleotide.

Conjugation The process whereby two bacteria exchange genetic material via a transient intercellular connection called a pilus.

Consensus sequence A nucleotide sequence used to describe a large number of related though non-identical sequences. Each position of the consensus sequence represents the nucleotide most often found at that position in the real sequences.

Continuous culture The culture of microorganisms in liquid medium under controlled conditions, with additions to and removals from the medium over a lengthy period of time.

Copy number The number of molecules of a plasmid contained in a single cell.

***Cos* site** One of the cohesive, single-stranded extensions present at the ends of the DNA molecules of certain strains of λ phage.

Cosmid A cloning vector consisting of the λ *cos* site inserted into a plasmid, used to clone DNA fragments up to 40 kb in size.

Covalently closed-circular (CCC)   A completely double-stranded circular DNA molecule, with no nicks or discontinuities, usually with a supercoiled conformation.

Defined medium   A bacterial growth medium in which all the components are known.

Deletion analysis   The identification of control sequences for a gene by determining the effects on gene expression of specific deletions in the upstream region.

Denaturation   Of nucleic acid molecules: breakdown by chemical or physical means of the hydrogen bonds involved in base pairing.

Density-gradient centrifugation   Separation of molecules and particles on the basis of buoyant density, by centrifugation in a concentrated sucrose or caesium chloride solution.

Dideoxynucleotide   A modified nucleotide that lacks the 3′ hydroxyl group and so prevents further chain elongation when incorporated into a growing polynucleotide.

Disarmed plasmid   A Ti plasmid that has had some or all of the T-DNA genes removed, so it is no longer able to promote cancerous growth of plant cells.

DNA sequencing   Determination of the order of nucleotides in a DNA molecule.

Double digestion   Cleavage of a DNA molecule with two different restriction endonucleases, either concurrently or consecutively.

Electrophoresis   Separation of molecules on the basis of their net electric charge.

Electroporation   A method for increasing DNA uptake by protoplasts through prior exposure to a high voltage which results in the temporary formation of small pores in the cell membrane.

End-filling   Conversion of a sticky end to a blunt end by enzymatic synthesis of the complement to the single-stranded extension.

Endonuclease   An enzyme that breaks phosphodiester bonds within a nucleic acid molecule.

Episome   A plasmid capable of integration into the host cell's chromosome.

Ethanol precipitation   Precipitation of nucleic acid molecules by ethanol plus salt, used primarily as a means of concentrating DNA.

Ethidium bromide   A fluorescent chemical that intercalates between base pairs in a double-stranded DNA molecule, used in the detection of DNA.

Exonuclease   An enzyme that sequentially removes nucleotides from the ends of a nucleic acid molecule.

Expression vector   A cloning vector designed so that a foreign gene inserted into the vector will be expressed in the host organism.

Fermenter   A vessel used for the large-scale culture of microorganisms.

Footprinting   The identification of protein-binding site on a DNA molecule by determining which phosphodiester bonds are protected from cleavage by DNase I.

Gel electrophoresis   Electrophoresis performed in a gel matrix so that molecules of similar electric charge can be separated on the basis of size.

Gel retardation   A technique that identifies a DNA fragment that has a bound protein by virtue of its decreased mobility during gel electrophoresis.

Geminivirus   One of the two groups of DNA viruses that infect plants, the members of which have potential as cloning vectors for some species of higher plants.

Gene   A segment of DNA that codes for an RNA and/or polypeptide molecule.

Gene cloning   Insertion of a fragment of DNA, carrying a gene, into a cloning vector, and subsequent propagation of the recombinant DNA molecule in a host organism.

Gene mapping   Determination of the relative positions of different genes on a DNA molecule.

Genetic engineering   The use of experimental techniques to produce DNA molecules containing new genes or new combinations of genes.

Genetics   The branch of biology devoted to the study of genes.

Genome   The complete set of genes of an organism.

Genomic library   A collection of clones sufficient in number to include all the genes of a particular organism.

Harvesting   The removal of microorganisms from a culture, usually by centrifugation.

Helper phage   A phage that is introduced into a host cell in conjunction with a related cloning vector, in order to provide enzymes required for replication of the cloning vector.

Heterologous probing   The use of a labelled nucleic acid molecule to identify related molecules by hybridization probing.

Homology   The degree of identity displayed by the nucleotide sequences of two related but not complementary polynucleotides. 85% homology means that 85 nucleotide positions out of 100 are identical in the two polynucleotides.

Homopolymer tailing   The attachment of a sequence of identical nucleotides (e.g. AAAAA) to the end of a nucleic acid molecule, usually referring to the synthesis of single-stranded homopolymer extensions on the ends of a double-stranded DNA molecule.

Horseradish peroxidase   An enzyme that can be complexed to DNA and which is used in a non-radioactive procedure for DNA labelling.

Host-controlled restriction   A mechanism by which some bacteria prevent phage attack through the synthesis of a restriction endonuclease that cleaves the non-bacterial DNA.

Hybrid-arrest translation (HART)   A method used to identify the polypeptide coded by a cloned gene.

Hybrid-release translation (HRT)   A method used to identify the polypeptide coded by a cloned gene.

Hybridization probe   A labelled nucleic acid molecule that can be used to identify complementary or homologous molecules through the formation of stable base-paired hybrids.

Immunological screening   The use of an antibody to detect a polypeptide synthesized by a cloned gene.

Incompatibility group   Comprises a number of different types of plasmid, often related to each other, that are unable to coexist in the same cell.

Induction   (1) Of a gene: the switching on of an expression of a gene or group of genes in response to a chemical or other stimulus. (2) Of λ phage: excision of the integrated form of λ, and switch to the lytic mode of infection, in response to a chemical or other stimulus.

Insertional inactivation   The cloning strategy whereby insertion of a new piece of DNA into a vector inactivates a gene carried by the vector.

Insertion vector   A λ vector constructed by deleting a segment of non-essential DNA.

***In situ* hybridization**   A technique for gene mapping involving hybridization of a labelled sample of a cloned gene to a large DNA molecule, usually a chromosome.

In vitro **mutagenesis** Any one of several techniques used to produce a specified mutation at a predetermined position in a DNA molecule.

In vitro **packaging** Synthesis of infective λ particles from a preparation of λ capsid proteins and a concatamer of DNA molecules separated by *cos* sites.

Klenow fragment (of DNA polymerase I) The enzyme that synthesizes a new DNA strand on an existing template, used primarily in DNA sequencing.

Labelling The incorporation of a radioactive nucleotide into a nucleic acid molecule.

Lambda (λ) A bacteriophage that infects *E. coli*, derivatives of which are extensively used as cloning vectors.

Ligase (DNA ligase) An enzyme that repairs single-stranded discontinuities in double-stranded DNA molecules in the cell. Purified DNA ligase is used in gene cloning to join DNA molecules together.

Linker A synthetic, double-stranded oligonucleotide used to attach sticky ends to a blunt-ended molecule.

Lysogen A bacterium that harbours a prophage.

Lysogenic infection cycle The pattern of phage infection that involves integration of the phage DNA into the host chromosome.

Lysozyme An enzyme that weakens the cell walls of certain types of bacteria.

Lytic infection cycle The pattern of infection displayed by a phage that replicates and lyses the host cell immediately after the initial infection. Integration of the phage DNA molecule into the bacterial chromosome does not occur.

M13 A bacteriophage that infects *E. coli*, derivatives of which are extensively used as cloning vectors.

Microinjection A method of introducing new DNA into a cell by injecting it directly into the nucleus.

Minimal medium A defined medium that provides only the minimum number of different nutrients needed for growth of a particular bacterium.

Multicopy plasmid A plasmid with a high copy number.

Multigene family A number of identical or related genes present in the same organism, usually coding for a family of related polypeptides.

Nick A single-strand break, involving the absence of one or more nucleotides, in a double-stranded DNA molecule.

Nick translation The repair of a nick with DNA polymerase I, usually to introduce labelled nucleotides into a DNA molecule.

Northern transfer A technique for transferring bands of RNA from an agarose gel to a nitrocellulose or similar membrane.

Nucleic acid hybridization Formation of a double-stranded molecule by base pairing between complementary or homologous polynucleotides.

Oligonucleotide-directed mutagenesis An *in vitro* mutagenesis technique that involves the use of a synthetic oligonucleotide to introduce the predetermined nucleotide alteration into the gene to be mutated.

Open-circular The non-supercoiled conformation taken up by a circular double-stranded DNA molecule when one or both polynucleotides carry nicks.

Origin of replication The specific position on a DNA molecule where DNA replication begins.

Orthogonal field alternation gel electrophoresis (OFAGE) A gel electrophoresis technique that employs a pulsed electric field to achieve separation of very large molecules of DNA.

Partial digestion Treatment of a DNA molecule with a restriction endonuclease under such conditions that only a fraction of all the recognition sites are cleaved.

Phenotypic expression   A technique designed to maximize the transformation frequency obtained when using a plasmid vector.

Pilus   One of the structures present on the surface of a bacterium containing a conjugative plasmid, through which DNA transfer occurs during conjugation.

Plaque   A zone of clearing on a lawn of bacteria caused by lysis of the cells by infecting phage particles.

Plasmid   A usually circular piece of DNA, primarily independent of the host chromosome, often found in bacterial and some other types of cell.

Plasmid amplification   A method involving incubation with an inhibitor of protein synthesis aimed at increasing the copy number of certain types of plasmid in a bacterial culture.

Polyethylene glycol   A polymeric compound used to precipitate macromolecules and molecular aggregates.

Polylinker   A synthetic double-stranded oligonucleotide carrying a number of restriction sites.

Primer   A short single-stranded oligonucleotide which, when attached by base pairing to a single-stranded template molecule, acts as the start point for complementary strand synthesis directed by a DNA polymerase enzyme.

Promoter   The nucleotide sequence, upstream of a gene, that acts as a signal for RNA polymerase binding.

Prophage   The integrated form of the DNA molecule of a lysogenic phage.

Protease   An enzyme that degrades protein.

Protein A   A protein from the bacterium *Staphylococcus aureus* that binds specifically to immunoglobulin G (i.e. antibody) molecules.

Protoplast   A cell from which the cell wall has been completely removed.

Radioactive marker   A radioactive atom used in the detection of a larger molecule in which it is incorporated.

Random priming   A method for DNA labelling that utilizes random DNA hexamers which will anneal to single-stranded DNA and act as primers for complementary-strand synthesis by a suitable enzyme.

Recombinant   A transformed cell that contains a recombinant DNA molecule.

Recombinant DNA molecule   A DNA molecule created in the test-tube by ligating together pieces of DNA that are not normally contiguous.

Recombinant DNA technology   All the techniques involved in the construction, study and use of recombinant DNA molecules.

Recombinant protein   A polypeptide that is synthesized in a recombinant cell as the result of expression of a cloned gene.

Recombination   The exchange of DNA sequences between different molecules, occurring either naturally or as a result of DNA manipulation.

Relaxed   Refers to the non-supercoiled conformation of open-circular DNA.

Replacement vector   A λ vector designed so that insertion of new DNA is by replacement of part of the non-essential region of the λ DNA molecule.

Replica plating   A technique whereby the colonies on an agar plate are transferred *en masse* to a new plate, on which the colonies will grow in the same relative positions as before.

Replicative form of M13   The double-stranded form of the M13 DNA molecule found within infected *E. coli* cells.

Reporter gene   A gene whose phenotype can be assayed in a transformed organism, and which is used in, for example, deletion analyses of regulatory regions.

Repression The switching off of expression of a gene or a group of genes in response to a chemical or other stimulus.

Restriction analysis Determination of the number and sizes of the DNA fragments produced when a particular DNA molecule is cut with a particular restriction endonuclease.

Restriction endonuclease An endonuclease that cuts DNA molecules only at a limited number of specific nucleotide sequences.

Restriction fragment length polymorphism (RFLP) A mutation that results in a detectable change in the pattern of fragments obtained when a DNA molecule is cut with a restriction endonuclease.

Restriction map A map showing the positions of different restriction sites in a DNA molecule.

RFLP linkage analysis A technique that uses a closely linked RFLP as a marker for the presence of a particular allele in a DNA sample, usually as a means of screening individuals for defective genes responsible for genetic diseases.

Ribonuclease An enzyme that degrades DNA.

Ribosome binding site The short nucleotide sequence upstream of a gene, which after transcription forms the site on the mRNA molecule to which the ribosome binds.

Selectable marker A gene carried by a vector and conferring a recognizable characteristic on a cell containing the vector or a recombinant DNA molecule derived from it.

Selection A means of obtaining a clone containing a desired recombinant DNA molecule.

Shotgun cloning A cloning strategy that involves the insertion of random fragments of a large DNA molecule into a vector, resulting in a large number of different recombinant DNA molecules.

Shuttle vector A vector that can replicate in the cells of more than one organism (e.g. in *E. coli* and in yeast).

Simian virus 40 (SV40) A mammalian virus used as the basis for a series of cloning vectors.

Southern transfer A technique for transferring bands of DNA from an agarose gel to a nitrocellulose or similar membrane.

Sphaeroplast A cell with a partially degraded cell wall.

Stem–loop A hairpin structure, consisting of a base-paired stem and a non-base-paired loop, that may form in a polynucleotide.

Sticky end An end of a double-stranded DNA molecule where there is a single-stranded extension.

Strong promoter An efficient promoter that can direct synthesis of RNA transcripts at a relatively fast rate.

Stuffer fragment The part of a λ replacement vector that is removed during insertion of new DNA.

Supercoiled The conformation of a covalently closed-circular DNA molecule, which is coiled by torsional strain into the shape taken by a wound-up elastic band.

T-DNA The portion of the Ti plasmid transferred to the plant DNA.

Temperature-sensitive mutation A mutation that results in a gene product that is functional within a certain temperature range (e.g. at less than 30°C), but non-functional at different temperatures (e.g. above 30°C).

Template A single-stranded polynucleotide (or region of a polynucleotide) able to direct synthesis of a complementary polynucleotide.

Terminator The short nucleotide sequence downstream of a gene that acts as a signal for termination of transcription.

5'-terminus One of the two ends of a polynucleotide; that which carries the phosphate group attached to the 5' position of the sugar.

3'-terminus One of the two ends of a polynucleotide; that which carries the hydroxyl group attached to the 3' position of the sugar.

Ti plasmid The large plasmid found in those *Agrobacterium tumefaciens* cells able to direct crown gall formation on certain species of plants.

Total cell DNA Consists of all the DNA present in a single cell or group of cells.

Transcript analysis Experiment aimed at determining which portions of a DNA molecule are transcribed into RNA.

Transfection The introduction of purified phage DNA molecules into a bacterial cell.

Transformation The introduction of any DNA molecule into any living cell.

Transformation frequency A measure of the proportion of cells in a population that are transformed in a single experiment.

Undefined medium A growth medium in which not all the components have been identified.

UV absorbance spectroscopy A method for measuring the concentration of a compound by determining the amount of ultraviolet radiation absorbed by a sample.

Vector A DNA molecule, capable of replication in a host organism, into which a gene is inserted to construct a recombinant DNA molecule.

Vehicle Often used as a substitute for the word 'vector', emphasizing that the vector transports the inserted gene through the cloning experiment.

Watson–Crick rules The base pairing rules that underlie gene structure and expression. A pairs with T, G with C.

Western transfer A technique for transferring bands of protein from an electrophoresis gel to a membrane support.

Yeast artificial chromosome (YAC) A cloning vector comprising the structural components of a yeast chromosome and able to clone very large pieces of DNA.

Yeast episomal plasmid (YEp) A yeast vector carrying the 2 μm circle origin of replication.

Yeast integrative plasmid (YIp) A yeast vector that relies on integration into the host chromosome for replication.

Yeast replicative plasmid (YRp) A yeast vector that carries a chromosomal origin of replication.

Index

(f = figure, t = table)